P9-BVG-863

TO
ENGINEER
IS HUMAN

TO ENGINEER IS HUMAN

The Role of Failure
in Successful Design

HENRY PETROSKI

ST. MARTIN'S PRESS/NEW YORK

'

Lines from "The Road Not Taken" from *The Poetry of Robert Frost* edited by Edward Connery Lathem. Copyright 1916, © 1969 by Holt, Rinehart and Winston. Copyright 1944 by Robert Frost. Reprinted by permission of Holt, Rinehart and Winston, publishers.

Lines from "Connoisseur of Chaos" from *The Collected Poems of Wallace Stevens.* Copyright © 1954 by Wallace Stevens. Reprinted by permission of Alfred A. Knopf, Inc., publishers.

Paragraphs from *The Memoirs of Herbert Hoover: Years of Adventure, 1874-1920.* Copyright © 1951 by Herbert Hoover. Reprinted by permission of the Hoover Foundation, New York, N. Y.

ILLUSTRATION CREDITS
1. *Punch.* 2. Reprinted with special permission, © 1979 by *The Philadelphia Inquirer/Washington Post* Writers Group. 3. Drawing by Chas. Addams, © 1984 by *The New Yorker* Magazine, Inc. 4. *Principles of the Mechanics of Machinery and Engineering,* by Julius Weisbach (Philadelphia: Lea and Blanchard, 1848). 5. *Dialogues Concerning Two New Sciences,* by Galileo Galilei (New York: Dover Publications, Inc., reprint of 1914 edition). 6. Adapted from *Elasticity in Engineering Mechanics,* by Arthur P. Boresi and Paul P. Lynn (Englewood Cliffs, N. J.: Prentice-Hall, Inc., 1974). 7. New York Historical Society. 8. Culver Pictures, Inc. 9. *Frank Leslie's Illustrated Newspaper.* 10. Victoria and Albert Museum. 11. *Illustrated London News.* 12. *Punch.* 13. Victoria and Albert Museum. 14. Courtesy Infomart. 15. Courtesy I. M. Pei & Partners, Architects and Planners. 16. Library of the Royal Institute of British Architects. 17. University of Washington. 18. Courtesy Michigan Department of Commerce. 19. Courtesy Freeman Fox, Consulting Engineers. 20. Reprinted with special permission from *Kansas City Star* Company, © 1981 by *Kansas City Star* Company. 21. Courtesy National Bureau of Standards. 22, 23, 24. The Bettmann Archive/UPI. 25. Frank Deceebo and Wayne Ratzenberger.

Some of this material has appeared, often in somewhat different form, in *Technology and Culture, Technology Review,* and *The Washington Post.*

Design by Barbara Richer

Library of Congress Cataloging in Publication Data
Petroski, Henry.
    To engineer is human.
    1. Engineering design. 2. System failures
(Engineering) I. Title.
TA174.P474 1985   620'.00425   85–1767
ISBN 0-312-80680-9

10 9 8 7 6 5 4

To Catherine

CONTENTS

List of Illustrations ix

Preface xi

1. Being Human 1

2. Falling Down Is Part of Growing Up 11

3. Lessons from Play; Lessons from Life 21
Appendix: "The Deacon's Masterpiece,"
    by Oliver Wendell Holmes 35

4. Engineering as Hypothesis 40

5. Success Is Foreseeing Failure 53

6. Design Is Getting From Here to There 64

7. Design as Revision 75

8. Accidents Waiting to Happen 85

9. Safety in Numbers 98

10. When Cracks Become Breakthroughs   107

11. Of Bus Frames and Knife Blades   122

12. Interlude: The Success Story of the
Crystal Palace   136

13. The Ups and Downs of Bridges   158

14. Forensic Engineering and Engineering
Fiction   172

15. From Slide Rule to Computer: Forgetting
How It Used to Be Done   189

16. Connoisseurs of Chaos   204

17. The Limits of Design   216

Bibliography   229

Index   241

LIST OF ILLUSTRATIONS

I. Cartoons illustrating public concern over engineering failures

II. Models of the ubiquitous cantilever beam

III. The Brooklyn Bridge: Anticipating failure by the engineer and by the layman

IV. The Crystal Palace: Testing the galleries and finding them sound

V. The Crystal Palace and two of its modern imitators

VI. Suspension bridges: The Tacoma Narrows and after

VII. The Kansas City Hyatt Regency walkways collapse

VIII. The Mianus River Bridge collapse and its aftermath

An eight-page section of illustrations follows after page 106.

PREFACE

Though ours is an age of high technology, the essence of what engineering is and what engineers do is not common knowledge. Even the most elementary of principles upon which great bridges, jumbo jets, or super computers are built are alien concepts to many. This is so in part because engineering as a human endeavor is not yet integrated into our culture and intellectual tradition. And while educators are currently wrestling with the problem of introducing technology into conventional academic curricula, thus better preparing today's students for life in a world increasingly technological, there is as yet no consensus as to how technological literacy can best be achieved.

I believe, and I argue in this essay, that the ideas of engineering are in fact in our bones and part of our human nature and experience. Furthermore, I believe that an understanding and an appreciation of engineers and engineering can be gotten without an engineering or technical education. Thus I hope that the technologically uninitiated will come to read what I have written as an introduction to technology. Indeed, this book is my answer to the questions "What is engineering?" and "What do engineers do?"

The idea of design—of making something that has not existed before—is central to engineering, and I take design and engineering to be virtually synonymous for the purposes of my development. Examples from structural designs commonly associated with mechanical and civil engineers are most prominent in this book because it is from those fields that I draw my own experiences, but the underlying principles are no less applicable to other branches of engineering.

I believe that the concept of failure—mechanical and structural failure in the context of this discussion—is central to understanding engineering, for engineering design has as its first and foremost objective the obviation of failure. Thus the colossal disasters that do occur are ultimately failures of design, but the lessons learned from those disasters can do more to advance engineering knowledge than all the successful machines and structures in the world. Indeed, failures appear to be inevitable in the wake of prolonged success, which encourages lower margins of safety. Failures in turn lead to greater safety margins and, hence, new periods of success. To understand what engineering is and what engineers do is to understand how failures can happen and how they can contribute more than successes to advance technology.

Any failures this book itself may have are surely of my own making, but I must recognize those works and people that gave me food for thought. The ambience of Duke University has been nurturing, and I have thoroughly enjoyed the opportunity it has given me to engage in both engineering research and in interdisciplinary programs with my colleagues in the School of Engineering, in Trinity College of Arts and Sciences, and in the Program in Science, Technology, and Human Values, in which both faculties come together. This broad spectrum of interactions has helped to widen my perspective.

The bibliography in this book is my implicit acknowledgment of the many places I have found support for my thesis about the role of failure in engineering design. Many of the more obscure documents I used were tracked down for me by Eric Smith, Duke's indefatigable engineering librarian. I have also benefitted from the term papers on case studies of structural failures that students prepared for the course in fracture mechanics and fatigue that I teach in the School of Engineering at Duke. My brother, William Petroski, a civil engineer, has been a continual source of information and opinion on structural failures, and on my visits with him, he has shown me many practical examples.

Certain physical arrangements enabled me to work on my manuscript without distractions and with modern tools. Albert Nelius has continued to understand my need for a carrel in Perkins Library, and for it I am grateful. My wife, Catherine Petroski, first encouraged me to use her word processor and has continued to make it available to me. I am fortunate that she is a day- and I a night-writer, and that this machine works wonderfully for both her fiction and my nonfiction and does not tire of either of our visions and revisions.

Several editors have encouraged me to write more and more ambitious pieces over the years, and I shall be forever grateful for their interest in my work. All the editors I have worked with at *Technology Review* have been a constant source of energy for me, and I am especially indebted to John Mattill, Tom Burroughs, and Steve Marcus, now at *High Technology,* for welcoming my contributions. Indeed, it was principally from the articles that Steve Marcus encouraged me to write for *Technology Review* that the present book has grown. And I am grateful to Tom Dunne of St. Martin's Press for giving me the opportunity to expand my ideas into a book.

My children, Karen and Stephen, with their questions and play, have enabled me to see the engineer in all of us, as their presence in this book will testify. And Catherine, who has from the very first demonstrated to me that no engineering concept is inaccessible to the English major, has shown me by example what it means to be a writer.

—Henry Petroski
Durham, North Carolina
September 1984

TO
ENGINEER
IS HUMAN

1
BEING HUMAN

Shortly after the Kansas City Hyatt Regency Hotel skywalks collapsed in 1981, one of my neighbors asked me how such a thing could happen. He wondered, did engineers not even know enough to build so simple a structure as an elevated walkway? He also recited to me the Tacoma Narrows Bridge collapse, the American Airlines DC–10 crash in Chicago, and other famous failures, throwing in a few things he had heard about hypothetical nuclear power plant accidents that were sure to exceed Three Mile Island in radiation release, as if to present an open-and-shut case that engineers did not quite have the world of their making under control.

I told my neighbor that predicting the strength and behavior of engineering structures is not always so simple and well-defined an undertaking as it might at first seem, but I do not think that I changed his mind about anything with my abstract generalizations and vague apologies. As I left him tending his vegetable garden and continued my walk toward home, I admitted to myself that I had not answered his question because I had not conveyed to him what engineering is. Without doing that I could not hope to explain what could go wrong with the products of engineering. In the years since the Hyatt Regency disaster I have thought a great deal about how I might explain the next technological embarrassment to an inquiring layman, and I have looked for examples not in the esoteric but in the commonplace. But I have also learned that collections of examples, no matter how vivid, no more make an explanation than do piles of beams and girders make a bridge.

Engineering has as its principal object not the given world but the world that engineers themselves create. And that world does not have the constancy of a honeycomb's design, changeless through countless generations of honeybees, for human structures involve constant and rapid evolution. It is not simply that we like change for the sake of change, though some may say that is reason enough. It is that human tastes, resources, and ambitions do not stay constant. We humans like our structures to be as fashionable as our art; we like extravagance when we are well off, and we grudgingly economize when times are not so good. And we like bigger, taller, longer things in ways that honeybees do not or cannot. All of these extra-engineering considerations make the task of the engineer perhaps more exciting and certainly less routine than that of an insect. But this constant change also introduces many more aspects to the design and analysis of engineering structures than there are in the structures of unimproved nature, and constant change means that there are many more ways in which something can go wrong.

Engineering is a human endeavor and thus it is subject to error. Some engineering errors are merely annoying, as when a new concrete building develops cracks that blemish it as it settles; some errors seem humanly unforgivable, as when a bridge collapses and causes the death of those who had taken its soundness for granted. Each age has had its share of technological annoyances and structural disasters, and one would think engineers might have learned by now from their mistakes how to avoid them. But recent years have seen some of the most costly structural accidents in terms of human life, misery, and anxiety, so that the record presents a confusing image of technological advancement that may cause some to ask, "Where is our progress?"

Any popular list of technological horror stories usually comprises the latest examples of accidents, failures, and flawed products. This catalog changes constantly as new disasters displace the old, but almost any list is representative of how varied the list itself can be. In 1979, when accidents seemed to be occurring left and right, anyone could rattle off a number of technological embarrassments that were fresh in everyone's mind, and there was no need to refer to old examples like the Tacoma Narrows Bridge to make the point. It seemed technology was running amok, and editorial pages across the country were anticipating the damage that might occur as the orbiting eighty-five-ton Skylab made its unplanned reentry. Many of the same newspapers also carried the cartoonist Tony Auth's solution to the problem. His cartoon shows the falling Skylab striking a flying DC–10, itself loaded with Ford Pintos fitted with Firestone 500 tires, with the entire wreckage falling on Three Mile Island, where the fire would be extinguished with asbestos hair dryers.

While such a variety may be unique to our times, the failure of the products of engineering is not. Almost four thousand years ago a number of Babylonian legal decisions were collected in what has come to be known as the Code of Hammurabi, after the sixth ruler of the First Dynasty of Babylon. There among nearly three hundred ancient cuneiform inscriptions governing matters like the status of women and drinking-house regulations are several that relate directly to the construction of dwellings and the responsibility for their safety:

If a builder build a house for a man and do not make its construction firm, and the house which he has built collapse and cause the death of the owner of the house, that builder shall be put to death.

If it cause the death of the son of the owner of the house, they shall put to death a son of that builder.

If it cause the death of a slave of the owner of the house, he shall give to the owner of the house a slave of equal value.

If it destroy property, he shall restore whatever it destroyed, and because he did not make the house which he built firm and it collapsed, he shall rebuild the house which collapsed from his own property.

If a builder build a house for a man and do not make its construction meet the requirements and a wall fall in, that builder shall strengthen the wall at his own expense.

This is a far cry from what happened in the wake of the collapse of the Hyatt Regency walkways, subsequently found to be far weaker than the Kansas City Building Code required. Amid a tangle of expert opinions, $3 billion in lawsuits were filed in the months after the collapse of the skywalks. Persons in the hotel the night of the accident were later offered $1,000 to sign on the dotted line, waiving all subsequent claims against the builder, the hotel, or anyone else they might have sued. And today opinions as to guilt or innocence in the Hyatt accident remain far from unanimous. After twenty months of investigation, the U. S. attorney and the Jackson County, Missouri, prosecutor jointly announced that they had found no evidence that a crime had been committed in connection with the accident. The attorney general of Missouri saw it differently, however, and he charged the engineers with "gross negligence." The engineers involved stand to lose their professional licenses but not their lives, but the verdict is still not in as I write three years after the accident.

The Kansas City tragedy was front-page news because it represented the largest loss of life from a building collapse in the history of the United States. The fact that it was news attests to the fact that countless buildings and structures, many with designs no less unique or daring than that of the hotel, are unremarkably safe. Estimates of the probability that a particular reinforced concrete or steel building in a technologically advanced country like the United States or England will fail in a given year range from one in a million to one in a hundred trillion, and the probability of death from a structural failure is approximately one in ten million per year. This is equivalent to a total of about twenty-five deaths per year in the United States, so that 114 persons killed in one accident in Kansas City was indeed news.

Automobile accidents claim on the order of fifty thousand American lives per year, but so many of these fatalities occur one or two at a time that they fail to create a sensational impact on the public. It seems to be only over holiday weekends, when the cumulative number of individual auto deaths reaches into the hundreds, that we acknowledge the severity of this chronic risk in our society. Otherwise, if an auto accident makes the front page or the evening news it is generally because an unusually large number of people or a person of note is involved. While there may be an exception if the dog is famous, the old saying that "dog bites man" is not news but that "man bites dog" is, applies.

We are both fascinated by and uncomfortable with the unfamiliar. When it was a relatively new technology, many people eschewed air travel for fear of a crash. Even now, when aviation relies on a well-established technology, many adults who do not think twice about the risks of driving an automobile are apprehensive about flying. They tell each other old jokes about white-knuckle air travelers, but younger generations who have come to use the airplane as naturally as their parents used the railroad and the automobile do not get the joke. Theirs is the rational attitude, for air travel *is* safe, the 1979 DC–10 crash in Chicago notwithstanding. Two years after that accident, the Federal Aviation Administration was able to announce that in the period covering 1980 and 1981, domestic airlines operated without a single fatal accident involving a large passenger jet. During the period of record, over half a billion passengers flew on ten million flights. Experience has proven that the risks of technology are very controllable.

However, as wars make clear, government administrations value their fiscal and political health as well as the lives of their citizens, and sometimes these objectives can be in conflict. The risks that engineered structures pose to human life and environments pose to society often conflict with the risks to the economy that striving for absolute and perfect safety would bring. We all know and daily make the trade-offs between our own lives and our pocketbooks, such as when we drive economy-sized automobiles that are incontrovertibly less safe than heavier-built ones. The introduction of seat belts, impact-absorbing bumpers, and emission-control devices have contributed to reducing risks, but gains like these have been achieved at a price to the consumer. Further improvements will take more time to perfect and will add still more to the price of a car, as the development of the air bag system has demonstrated. Thus there is a constant tension between manufacturers and consumer advocates to produce safe cars at reasonable prices.

So it is with engineering and public safety. All bridges and buildings could be built ten times as strong as they presently are, but at a tremendous increase in cost, whether financed by taxes or private investment. And, it would be argued, why ten times stronger? Since so few bridges and buildings collapse now, surely ten times stronger would be structural overkill. Such ultraconservatism would strain our economy and make our built environment so bulky and massive that architecture and style as we know them would have to undergo radical change. No, it would be argued, ten times is too much stronger. How about five? But five might also arguably be considered too strong, and a haggling over numbers representing no change from the present specifications and those representing five- or a thousand-percent improvement in strength might go on for as long as Zeno imagined it would take him to get from here to there. But less-developed countries may not have the luxury to argue about risk or debate paradoxes, and thus their buildings and boilers can be expected to collapse and explode with what appears to us to be uncommon frequency.

Callous though it may seem, the effects of structural reliability can be measured not only in terms of cost in human lives but also in material terms. This was done in a recent study conducted by the National Bureau of Standards with the assistance of Battelle Columbus Laboratories. The study found that fracture, which included such diverse phenomena as the breaking of eyeglasses, the cracking of highway pavement, the collapse of bridges, and the breakdown of machinery, costs well over $100 billion annually, not only for actual but also for anticipated replacement of broken parts and for structural insurance against parts breaking in the first place. Primarily associated with the transportation and construction industries, many of these expenses arise through the prevention of fracture by overdesign (making things heavier than otherwise necessary) and maintenance (watching for cracks to develop), and through the capital equipment investment costs involved in keeping spare parts on hand in anticipation of failures. The 1983 report further concludes that the costs associated with fracture could be reduced by one half by our better utilizing available technology and by improved techniques of fracture control expected from future research and development.

Recent studies of the condition of our infrastructure—the water supply and sewer systems, and the networks of highways and bridges that we by and large take for granted—conclude that it has been so sorely neglected in many areas of the country that it would take billions upon billions of dollars to put things back in shape. (Some estimates put the total bill as high as $3 trillion.) This condition resulted in part from maintenance being put off to save money during years when energy and personnel costs were taking ever-larger slices of municipal budget pies. Some water pipes in large cities like New York are one hundred or more years old, and they were neither designed nor expected to last forever. Ideally, such pipes should be replaced on an ongoing basis to keep the whole water supply system in a reasonably sound condition, so that sudden water main breaks occur very infrequently. Such breaks can have staggering consequences, as when a main installed in 1915 broke in 1983 in midtown Manhattan and flooded an underground power station, causing a fire. The failure of six transformers interrupted electrical service for several days. These happened to be the same days of the year that ten thousand buyers from across the country visited New York's garment district to purchase the next season's lines. The area covered by the blackout just happened to be the blocks containing the showrooms of the clothing industry, so that there was mayhem where there would ordinarily have been only madness. Financial losses due to disrupted business were put in the millions.

In order to understand how engineers endeavor to insure against such structural, mechanical, and systems failures, and thereby also to understand how mistakes can be made and accidents with far-reaching consequences can occur, it is necessary to understand, at least partly, the nature of engineering design. It is the process of design, in which diverse parts of the "given-world" of the scientist and the "made-world" of the engineer are re-formed and assembled into something the likes of which Nature had not dreamed, that divorces engineering from science and marries it to art. While the practice of engineering may involve as much technical experience as the poet brings to the blank page, the painter to the empty canvas, or the composer to the silent keyboard, the understanding and appreciation of the process and products of engineering are no less accessible than a poem, a painting, or a piece of music. Indeed, just as we all have experienced the rudiments of artistic creativity in the childhood masterpieces our parents were so proud of, so we have all experienced the essence of structural engineering in our learning to balance first our bodies and later our blocks in ever more ambitious positions. We have learned to endure the most boring of cocktail parties without the social accident of either our bodies or our glasses succumbing to the force of gravity, having long ago learned to crawl, sit up, and toddle among our tottering towers of blocks. If we could remember those early efforts of ours to raise ourselves up among the towers of legs of our parents and their friends, then we can begin to appreciate the task and the achievements of engineers, whether they be called builders in Babylon or scientists in Los Alamos. For all of their efforts are to one end: to make something stand that has not stood before, to reassemble Nature into something new, and above all to obviate failure in the effort.

Because man is fallible, so are his constructions, however. Thus the history of structural engineering, indeed the history of engineering in general, may be told in its failures as well as in its triumphs. Success may be grand, but disappointment can often teach us more. It is for this reason that hardly a history can be written that does not include the classic blunders, which more often than not signal new beginnings and new triumphs. The Code of Hammurabi may have encouraged sound construction of reproducible dwellings, but it could not have encouraged the evolution of the house, not to mention the skyscraper and the bridge, for what builder would have found incentive in the code to build what he believed to be a better but untried house? This is not to say that engineers should be given license to experiment with abandon, but rather to recognize that human nature appears to want to go beyond the past, in building as in art, and that engineering is a human endeavor.

When I was a student of engineering I came to fear the responsibility that I imagined might befall me after graduation. How, I wondered, could I ever be perfectly sure that something I might design would not break or collapse and kill a number of people? I knew my understanding of my textbooks was less than total, my homework was seldom without some sort of error, and my grades were not straight As. This disturbed me for some time, and I wondered why my classmates, both the A and C students, were not immobilized by the same phobia. The topic never came to the surface of our conversations, however, and I avoided confronting the issue by going to graduate school instead of taking an engineering job right away. Since then I have come to realize that my concern was not unique among engineering students, and indeed many if not all students have experienced self-doubts about success and fears of failure. The medical student worries about losing a patient, the lawyer about losing a crucial case. But if we all were to retreat with our phobias from our respective jobs and professions, we could cause exactly what we wish to avoid. It is thus that we practice whatever we do with as much assiduousness as we can command, and we hope for the best. The rarity of structural failures attests to the fact that engineering at least, even at its most daring, is not inclined to take undue risks.

The question, then, should not only be why do structural accidents occur but also why not more of them? Statistics show the headline-grabbing failure to be as rare as its newsworthiness suggests it to be, but to understand why the risk of structural failure is not absolutely zero, we must understand the unique engineering problem of designing what has not existed before. By understanding this we will come to appreciate not only why the probability of failure is so low but also how difficult it might be to make it lower. While it is theoretically possible to make the number representing risk as close to zero as desired, human nature in its collective and individual manifestations seems to work against achieving such a risk-free society.

2
FALLING DOWN IS
PART OF GROWING UP

We are all engineers of sorts, for we all have the principles of machines and structures in our bones. We have learned to hold our bodies against the forces of nature as surely as we have learned to walk. We calculate the paths of our arms and legs with the computer of our brain, and we catch baseballs and footballs with more dependability than the most advanced weapons systems intercept missiles. We may wonder if human evolution may not have been the greatest engineering feat of all time. And though many of us forget how much we once knew about the principles and practice of engineering, the nursery rhymes and fairy tales of our youth preserve the evidence that we did know quite a bit.

We are born into a world swathed in trust and risk. And we become accustomed from the instant of birth to living with the simultaneous possibilities that there *will* be and that there will *not* be catastrophic structural failure. The doctor who delivers us and the nurses who carry us about the delivery room are cavalier human cranes and forklifts who have moved myriad babies from delivery to holding upside down to showing to mother to cleansing to footprinting to wristbanding to holding right-side up to showing to father to taking to the nursery. I watched with my heart in my mouth as my own children were so moved and rearranged, and the experience exhausted me. Surely sometime, somewhere, a baby has been dropped, surely a doctor has had butterfingers or a nurse a lapse of attention. But we as infants and we as parents cannot and do not and should not dwell on those remotely possible, hideous scenarios, or we might immobilize the human race in the delivery room. Instead, our nursery rhymes help us think about the unthinkable in terms of serenity.

> *Rock-a-bye baby*
> *In the tree top.*
> *When the wind blows,*
> *The cradle will rock.*
> *When the bough breaks,*
> *The cradle will fall.*
> *And down will come baby,*
> *Cradle and all.*

Home from the hospital, we are in the hands of our parents and friends and relatives—and structurally weak siblings. We are held up helpless over deep pile carpets and hard terrazzo floors alike, and we ride before we walk, risking the sudden collapse of an uncle's trick knee. We are transported across impromptu bridges of arms thrown up without plans or blueprints between mother and aunt, between neighbor and father, between brother and sister —none of whom is a registered structural engineer. We come to Mama and to Papa eventually to forget our scare reflex and we learn to trust the beams and girders and columns of their arms and our cribs. We become one with the world and nap in the lap of gravity. Our minds dream weightlessly, but our ears come to hear the sounds of waking up. We listen to the warm whispers giving structure to the world of silence, and we learn from the bridges of lullabyes and play that not only we but also the infrastructure needs attention.

London Bridge is falling down,
Falling down, falling down.
London Bridge is falling down,
My fair lady.

Build it up with wood and stone,
Wood and stone, wood and stone.
Build it up with wood and stone,
My fair lady.

The parts of our bodies learn to function as levers, beams, columns, and even structures like derricks and bridges as we learn to turn over in our cribs, to sit up, to crawl, to walk, and generally to support the weight of our own bodies as well as what we lift and carry. At first we do these things clumsily, but we learn from our mistakes. Each time the bridge of our body falls down, we build it up again. We pile back on hands and knees to crawl over the river meandering beneath us. We come to master crawling, and we come to elaborate upon it, moving faster and freer and with less and less concern for collapsing all loose in the beams and columns of our back and limbs. We extend our infant theory of structures and hypothesize that we can walk erect, cantilevering our semicircular canals in the stratosphere. We think these words in the Esperanto of babble, and with the arrogance of youth we reach for the stars. With each tottering attempt to walk, our bodies learn from the falls what not to do next time. In time we walk without thinking and think without falling, but it is not so much that we have learned how to walk as we have learned not to fall. Sometimes we have accidents and we break our arms and legs. We have them fixed and we go on as before. Barring disease, we walk erect and correctly throughout our lives until our structure deteriorates with old age and we need to be propped up with canes or the like. For the majority of our lives walking generally becomes as dependable as one can imagine it to be, but if we choose to load the structure of our bodies beyond the familiar limits of walking, say by jogging or marathoning, then we run the risk of structural failure in the form of muscle pulls and bone fractures. But our sense of pain stops most of us from overexerting ourselves and from coming loose at our connections as we go round and round, hand in hand, day in and day out.

> *Ring around the rosie,*
> *A pocket full of posies,*
> *Ashes, ashes,*
> *We all fall down.*

If ontogeny recapitulates phylogeny, if all that has come to be human races before the fetus floating in its own prehistory, then the child playing relives the evolution of structural engineering in its blocks. And the blocks will be as stone and will endure as monuments to childhood, as Erector Sets and Tinker Toys and Legos will not. Those modern optimizations will long have folded and snapped in the frames and bridges of experiment, though not before the child will have learned from them the limitations of metal and wood and plastic. These lessons will be carried in the tool box of the mind to serve the carpenter in all of us in time.

> *Step on a crack*
> *And break your mother's back.*

The child will play with mud and clay, making cakes and bricks in the wonderful oven of the sun. The child will learn that concrete cracks a mother's back but that children's backs are as resilient as springs and pliant as saplings. The child will watch the erection of flowers on columns of green but break them for the smiles of its parents. Summer will roof houses in the bushes, vault cathedrals in the trees. The child will learn the meaning of time, and watch the structures fall into winter and become skeletons of shelters that will be built again out of the dark in the ground and the light in the sky. The child angry and victimized by other children angry will learn the meanings of vandalism and sabotage, of demolition and destruction, of collapse and decline, of the lifetime of structures—and the structure of life.

The Sphinx asked, "What walks on four legs in the morning, two legs in the afternoon, and three legs in the evening?"

The child learns that the arms and legs of dolls and soldiers break, the wheels of wagons and tricycles turn against their purpose, and the bats and balls of games do not last forever. No child articulates it, but everyone learns that toys are mean. They teach us not the vocabulary but the reality of structural failure and product liability. They teach us that as we grow, the toys that we could not carry soon cannot carry us. They are as bridges built for the traffic of a lighter age, and their makers are as blameless as the builders of a lighter bridge. We learn that not everything can be fixed.

Humpty Dumpty sat on a wall;
Humpty Dumpty had a great fall.
All the King's horses and all the King's men
Couldn't put Humpty together again.

The adolescent learns that bones can break. The arms counterbalancing the legs locomoting are as fragile as the steel and iron railroad bridges under the reciprocating blows of the behemoths rushing through the nineteenth century. The cast of thousands of childhoods reminds the arms and legs, while they have grown stronger but brittler, that they have also grown taller and wiser. They fall less and less. They grow into the arms and legs of young adults making babies fly between them, wheeeee, up in the air unafraid of the gravity parents can throw away. But the weight of responsibility and bills and growing babies brings the parents down to earth and they begin to think of things besides their bridges of muscles and columns of bones. They think of jobs and joys of a different kind, perhaps even if they are engineers.

> *Jack and Jill went up the hill*
> *To fetch a pail of water,*
> *Jack fell down and broke his crown*
> *And Jill came tumbling after.*

The natural fragileness of things comes to be forgotten, for we have learned to take it easy on the man-made world. We do not pile too high or reach too far. We make our pencil points sharper, but we do not press as hard. We learn to write without snap, and the story of our life goes smoothly, but quickly becomes dull. (Everyone wishes secretly to be the writer pushing the pencil to its breaking point.) We feel it in our bones as we grow old and then we remember how brittle but exhilarating life can be. And we extend ourselves beyond our years and break our bones again, thinking what the hell. We have wisdom and we understand the odds and probabilities. We know that nothing is forever.

> *Three wise men of Gotham*
> *Went to sea in a bowl:*
> *If the vessel had been stronger,*
> *My song would be longer.*

As if it were not enough that the behavior of our very bodies accustoms us to the limitations of engineering structures, our language itself is ambiguous about the daily trials to which life and limb are subjected. Both human beings and inhuman beams are said to be under stress and strain that may lead to fatigue if not downright collapse. Breakdowns of man and machine can occur if they are called upon to carry more than they can bear. The anthropomorphic language of engineering is perhaps no accident since man is not only the archetypal machine but also the Ur-structure.

Furniture is among the oldest of inanimate engineering structures designed to carry a rather well-defined load under rather well-defined circumstances. We are not surprised that furniture used beyond its intended purpose is broken, and we readily blame the child who abuses the furniture rather than the designer of the furniture or the furniture itself when it is abused. Thus a chair must support a person in a sitting position, but it might not be expected to survive a brawl in a saloon. A bed might be expected to support a recumbent child, a small rocking chair only a toddler. But the child's bed would not necessarily be considered badly designed if it collapsed under the child's wild use of it as a trampoline, and a child's chair cannot be faulted for breaking under the weight of a heavier child using it as a springboard. The arms and legs of chairs, the heads and feet of beds, just like those of the people whom they serve, cannot be expected to be strong without limit.

Mother Goose is as full of structural failures as human history. The nursery rhymes acknowledge the limitations of the strength of the objects man builds as readily as fairy tales recognize the frailties of human nature. The story of Goldilocks and the Three Bears teaches us how we can unwittingly proceed from engineering success to failure. Papa Bear's chair is so large and so hard and so unyielding under the weight of Goldilocks that apparently without thinking she gains a confidence in the strength of all rocking chairs. Goldilocks next tries Mama Bear's chair, which is not so large but is softer, perhaps because it is built with a lighter wood. Goldilocks finds this chair too soft, however, too yielding in the cushion. Yet it is strong enough to support her. Thus the criterion of strength becomes less a matter of concern than the criteria of "give" and comfort, and Goldilocks is distracted by her quest for a comfortable chair at the expense of one sufficiently strong. Finally Goldilocks approaches Baby Bear's chair, which is apparently stiffer but weaker than Mama Bear's, with little if any apprehension about its safety, for Goldilocks' experience is that all chairs are overdesigned. At first the smallest chair appears to be "just right," but, as with all marginal engineering designs, whether chairs or elevated walkways, the chair suddenly gives way under Goldilocks and sends her crashing to the floor.

The failure of the chair does not keep Goldilocks from next trying beds without any apparent concern for their structural integrity. When Papa Bear's bed is too hard and Mama's is too soft, Goldilocks does not seem to draw a parallel with the chairs. She finds Baby Bear's bed "just right" and falls asleep in it without worrying about its collapsing under her. One thing the fairy tale implicitly teaches us as children is to live in a world of seemingly capricious structural failure and success without anxiety. While Goldilocks may worry about having broken Baby Bear's chair, she does not worry about all chairs and beds breaking. According to Bruno Bettelheim, the tale of Goldilocks and the Three Bears lacks some of the important features of a true fairy tale, for in it there is neither recovery nor consolation, there is no resolution of conflict, and Goldilocks' running away from the bears is not exactly a happy ending. Yet there is structural recovery and consolation in that the bed does not break, and there is thereby a structural happy ending.

If the story of Goldilocks demonstrates how the user of engineering products can be distracted into overestimating their strength, the story of the Three Little Pigs shows how the designer can underestimate the strength his structure may need in an emergency or, as modern euphemisms would put it, under extreme load or hypothetical accident conditions. We recall that each of the three pigs has the same objective: to build a house. It is implicit in the mother pig's admonishment as they set out that their houses not only will have to shelter the little pigs from ordinary weather, but must also stand up against any extremes to which the Big Bad Wolf may subject them.

The three little pigs are all aware of the structural requirements necessary to keep the wolf out, but they differ in their beliefs of how severe a wolf's onslaught can be, and some of the pigs would like to get by with the least work and the most play. Thus the individual pigs make different estimates of how strong their houses must be, and each reaches a different conclusion about how much strength he can sacrifice to availability of materials and time of construction. That each pig thinks he is building his house strong enough is demonstrated by the first two pigs dancing and singing, "Who's afraid of the Big Bad Wolf." They think their houses are safe enough and that their brother laboring over his brick house has overestimated the strength of the wolf and overdesigned his structure. Finally, when the third pig's house is completed, they all dance and sing their assurances. It is only the test of the wolf's full fury that ultimately proves the third pig correct. Had the wolf been a bugaboo, all three houses might have stood for many a year and the first two pigs never been proven wrong.

Thus the nursery rhymes, riddles, and fairy tales of childhood introduce us to engineering. From lullabyes that comfort us even as they sing of structural failure to fairy tales that teach us that we can build our structures so strong that they can withstand even the huffing and puffing of a Big Bad Wolf, we learn the rudiments and the humanness of engineering.

Our own bodies, the oral tradition of our language and our nursery rhymes, our experiences with blocks and sand, all serve to accustom us to the idea that structural failure is part of the human condition. Thus we seem to be preconditioned, or at least emotionally prepared, to expect bridges and dams, buildings and boats, to break now and then. But we seem not at all resigned to the idea of major engineering structures having the same mortality as we. Somehow, as adults who forget their childhood, we expect our constructions to have evolved into monuments, not into mistakes. It is as if engineers and non-engineers alike, being human, want their creations to be superhuman. And that may not seem to be an unrealistic aspiration, for the flesh and bone of steel and stone can seem immortal when compared with the likes of man.

3
LESSONS FROM PLAY; LESSONS FROM LIFE

When I want to introduce the engineering concept of fatigue to students, I bring a box of paper clips to class. In front of the class I open one of the paper clips flat and then bend it back and forth until it breaks in two. That, I tell the class, is failure by fatigue, and I point out that the number of back and forth cycles it takes to break the paper clip depends not only on how strong the clip is but also on how severely I bend it. When paper clips are used normally, to clip a few sheets of paper together, they can withstand perhaps thousands or millions of the slight openings and closings it takes to put them on and take them off the papers, and thus we seldom experience their breaking. But when paper clips are bent open so wide that they look as if we want them to hold all the pages of a book together, it might take only ten or twenty flexings to bring them to the point of separation.

Having said this, I pass out a half dozen or so clips to each of the students and ask them to bend their clips to breaking by flexing them as far open and as far closed as I did. As the students begin this low-budget experiment, I prepare at the blackboard to record how many back and forth bendings it takes to break each paper clip. As the students call out the numbers, I plot them on a bar graph called a histogram. Invariably the results fall clearly under a bell-shaped normal curve that indicates the statistical distribution of the results, and I elicit from the students the explanations as to why not all the paper clips broke with the same number of bendings. Everyone usually agrees on two main reasons: not all paper clips are equally strong, and not every student bends his clips in exactly the same way. Thus the students recognize at once the phenomenon of fatigue and the fact that failure by fatigue is not a precisely predictable event.

Many of the small annoyances of daily life are due to predictable—but not precisely so—fractures from repeated use. Shoelaces and light bulbs, as well as many other familiar objects, seem to fail us suddenly and when it is least convenient. They break and burn out under conditions that seem no more severe than those they had been subjected to hundreds or thousands of times before. A bulb that has burned continuously for decades may appear in a book of world records, but to an engineer versed in the phenomenon of fatigue, the performance is not remarkable. Only if the bulb had been turned on and off daily all those years would its endurance be extraordinary, for it is the cyclic and not the continuous heating of the filament that is its undoing. Thus, because of the fatiguing effect of being constantly changed, it is the rare scoreboard that does not have at least one bulb blown.

Children's toys are especially prone to fatigue failure, not only because children subject them to seemingly endless hours of use but also because the toys are generally not overdesigned. Building a toy too rugged could make it too heavy for the child to manipulate, not to mention more expensive than its imitators. Thus, the seams of rubber balls crack open after so many bounces, the joints of metal tricycles break after so many trips around the block, and the heads of plastic dolls separate after so many nods of agreement.

Even one of the most innovative electronic toys of recent years has been the victim of mechanical fatigue long before children (and their parents) tire of playing with it. Texas Instruments' Speak & Spell effectively employs one of the first microelectronic voice synthesizers. The bright red plastic toy asks the child in a now-familiar voice to spell a vocabulary of words from the toy's memory. The child pecks out letters on the keyboard, and they appear on a calculator-like display. When the child finishes spelling a word, the ENTER key is pressed and the computer toy says whether the spelling is correct and prompts the child to try again when a word is misspelled. Speak & Spell is so sophisticated that it will turn itself off if the child does not press a button for five minutes or so, thus conserving its four C-cells.

My son's early model Speak & Spell had given him what seemed to be hundreds of hours of enjoyment when one day the ENTER key broke off at its plastic hinge. But since Stephen could still fit his small finger into the buttonhole to activate the switch, he continued to enjoy the smart, if disfigured, toy. Soon thereafter, however, the E key snapped off, and soon the T and O keys followed suit. Although he continued to use the toy, its keyboard soon became a maze of missing letters and, for those that were saved from the vacuum cleaner, taped-on buttons.

What made these failures so interesting to me was the very strong correlation between the most frequently occurring letters in the English language and the fatigued keys on Stephen's Speak & Spell. It is not surprising that the ENTER key broke first, since it was employed for inputting each word and thus got more use than any one letter. Of the seven most common letters—in decreasing occurrence, E, T, A, O, I, N, S, R—five (E, T, O, S, and R) were among the first keys to break. All other letter keys, save for the two seemingly anomalous failures of P and Y, were intact when I first reported this serendipitous experiment on the fatigue phenomenon in the pages of *Technology Review*.

If one assumes that all Speak & Spell letter keys were made as equally well as manufacturing processes allowed, perhaps about as uniformly as or even more so than paper clips, then those plastic keys that failed must generally have been the ones pressed most frequently. The correlation between letter occurrence in common English words and the failure of the keys substantiates that this did indeed happen, for the anomalous failures seem also to be explainable in terms of abnormally high use. Because my son is right-handed, he might be expected to favor letters on the right-hand side of the keyboard when guessing spellings or just playing at pressing letters. Since none of the initial failed letters occurs in the four left-most columns of Speak & Spell, this proclivity could also explain why the common-letter keys A and N were still intact. The anomalous survival of the I key may be attributed to its statistically abnormal strength or to its underuse by a gregarious child. And the failure of the infrequently occurring P and Y might have been a manifestation of the statistical weakness of the keys or of their overuse by my son. His frequent spelling of his name and of the name of his cat, Pollux, endeared the letter P to him, and he had learned early that Y is sometimes a vowel. Furthermore, each time the Y key was pressed, Speak & Spell would ask the child's favorite question, "Why?"

Why the fatigue of its plastic buttons should have been the weak link that destroyed the integrity of my son's most modern electronic toy could represent the central question for understanding engineering design. Why did the designers of the toy apparently not anticipate this problem? Why did they not use buttons that would outlast the toy's electronics? Why did they not obviate the problem of fatigue, the problem that has defined the lifetimes of mechanical and structural designs for ages? Such questions are not unlike those that are asked after the collapse of a bridge or the crash of an airplane. But the collapse of a bridge or the crash of an airplane can endanger hundreds of lives, and thus the possibility of the fatigue of any part can be a lesson from which its victims learn nothing. Yet the failure of a child's toy, though it may cause tears, is but a lesson for a child's future of burnt-out light bulbs and broken shoelaces. And years later, when his shoelaces break as he is rushing to dress for an important appointment, he will be no less likely to ask, "Why?"

After I wrote about the found experiment, my son retrieved his

Speak & Spell from my desk and resumed playing with the toy—and so continued the experiment. Soon another key failed, the vowel key U in the lower left position near where Stephen held his thumb. Next the A key broke, another vowel and the third most frequently occurring letter of the alphabet. The experiment ended with that failure, however, for Stephen acquired a new model of Speak & Spell with the new keyboard design that my daughter, Karen, had pointed out to me at an electronics store. Instead of having individually hinged plastic buttons, the new model has its keyboard printed on a single piece of rubbery plastic stretched over the switches. The new model Stephen has is called an E. T. Speak & Spell, after the little alien creature in the movie, and I am watching the plastic sheet in the vicinity of those two most frequently occurring letters to see if the fatigue gremlin will strike again.

Not long after I had first written about my son's Speak & Spell I found out from readers that their children too had had to live with disfigured keyboards. It is a tribute to the ingeniousness of the toy—and the attachment that children had developed for it—that they endured the broken keys and adapted in makeshift ways, as they would have to throughout a life of breakdowns and failures in our less than perfect world. Some parents reported that their children apparently discovered that the eraser end of a pencil fit nicely into the holes of the old Speak & Spell and thus could be used to enter the most frequently used letters without the children having to use their fingertips. I have wondered if indeed this trick was actually discovered by the parents who loved to play with the toy, for almost any child's finger should easily fit into the hole left by the broken button, but Mommy or Daddy's certainly would not.

Nevertheless, this resourcefulness suggests that the toy would have been a commercial success even with its faults, but the company still improved the keyboard design to solve the problem of key fatigue. The new buttonless keyboard is easily cleaned and pressed by even the clumsiest of adult fingers. The evolution of the Speak & Spell keyboard is not an atypical example of the way mass-produced items, though not necessarily planned that way, are debugged through use. Although there may have been some disappointment among parents who had paid a considerable amount of money for what was then among the most advanced applications of microelectronics wizardry, their children, who were closer to the world of learning to walk and talk and who were still humbled by their skinned knees and twisted tongues, took the failure of the keys in stride. Perhaps the manufacturer of the toy, in the excitement of putting the first talking computer on the market, overlooked some of the more mundane aspects of its design, but when the problem of the fractured keys came to its attention, it acted quickly to improve the toy's mechanical short-comings.

I remember being rather angry when my son's Speak & Spell lost its first key. For all my understanding of the limitations of engineering and for all my attempted explanations to my neighbors of how failures like the Hyatt Regency walkways and the DC-10 could happen without clear culpability, I did not extend my charity to the designers of the toy. But there is a difference in the design and development of things that are produced by the millions and those that are unique, and it is generally the case that the mass-produced mechanical or electronic object undergoes some of its debugging and evolution after it is offered to the consumer. Such actions as producing a new version of a toy or carrying out an automobile recall campaign are not possible for the large civil engineering structure, however, which must be got right from the first stages of construction. So my charity should have extended to the designers of the Speak & Spell, for honest mistakes can be made by mechanical and electrical as well as by civil engineers. Perhaps someone had underestimated the number of Es it would take a child to become bored with the new toy. After all, most toys are put away long before they break. If this toy, which is more sophisticated than any I ever had in my own childhood, could tell me when I misspelled words I never could keep straight, then I would demand from it other superhuman qualities such as indestructibility. Yet we do not expect that of everything.

Although we might all be annoyed when a light bulb or a shoelace breaks, especially if it does so at a very inconvenient time, few if any of us would dream of taking it back to the store claiming it had malfunctioned. We all know the story of Thomas Edison searching for a suitable filament for the light bulb, and we are aware of and grateful for the technological achievement. We know, almost intuitively it seems, that to make a shoelace that would not break would involve compromises that we are not prepared to accept. Such a lace might be undesirably heavy or expensive for the style of shoe we wear, and we are much more willing to have the option of living with the risk of having the lace break at an inopportune time or of having the small mental burden of anticipating when the lace will break so that we might replace it in time. Unless we are uncommonly fastidious, we live dangerously and pay little attention to preventive maintenance of our fraying shoelaces or our aging light bulbs. Though we may still ask "Why?" when they break, we already know and accept the answer.

As the consequences of failure become more severe, however, the forethought we must give to them becomes more a matter of life and death. Automobiles are manufactured by the millions, but it would not do to have them failing with a snap on the highways the way light bulbs and shoelaces do at home. The way an automobile could fail must be anticipated so that, as much as possible, a malfunction does not lead to an otherwise avoidable deadly accident. Since tires are prone to flats, we want our vehicles to be able to be steered safely to the side of the road when one occurs. Such a failure is accepted in the way light bulb and shoelace failures are, and we carry a spare tire to deal with it. Other kinds of malfunctions are less acceptable. We do not want the brakes on all four wheels and the emergency braking system to fail us suddenly and simultaneously. We do not want the steering wheel to come off in our hands as we are negotiating a snaking mountain road. Certain parts of the automobile are given special attention, and in the rare instances when they do fail, leading to disaster, massive lawsuits can result. When they become aware of a potential hazard, automobile manufacturers are compelled to eliminate what might be the causes of even the most remote possibilities of design-related accidents by the massive recall campaigns familiar to us all.

As much as it is human to make mistakes, it is also human to want to avoid them. Murphy's Law, holding that anything that can go wrong will, is not a law of nature but a joke. All the light bulbs that last until we tire of the lamp, all the shoelaces that outlast their shoes, all the automobiles that give trouble-free service until they are traded in have the last laugh on Murphy. Just as he will not outlive his law, so nothing manufactured can be or is expected to last forever. Once we recognize this elementary fact, the possibility of a machine or a building being as near to perfect for its designed lifetime as its creators may strive to be for theirs is not only a realistic goal for engineers but also a reasonable expectation for consumers. It is only when we set ourselves such an unrealistic goal as buying a shoelace that will never break, inventing a perpetual motion machine, or building a vehicle that will never break down that we appear to be fools and not rational beings.

Oliver Wendell Holmes is remembered more widely for his humor and verse than for the study entitled "The Contagiousness of Puerperal Fever" that he carried out as Parkman Professor of Anatomy and Physiology at Harvard Medical School. Yet it may have been his understanding of the seemingly independent working of the various parts of the human body that helped him to translate his physiological experiences into a lesson for structural and mechanical engineers. Although some of us go first in the knees and others in the back, none of us falls apart all at once in all our joints. So Holmes imagined the foolishness of expecting to design a horse-drawn carriage that did not have a weak link.

Although intended as an attack on Calvinism, in which Holmes uses the metaphor of the "one-hoss shay" to show that a system of logic, no matter how perfect it seems, must collapse if its premises are false, the poem also holds up as a good lesson for engineers. Indeed, Micro-Measurements, a Raleigh, North Carolina-based supplier of devices to measure the stresses and strains in engineering machines and structures, thinks "The Deacon's Masterpiece" so apt to its business that it offers copies of the poem suitable for framing. The firm's advertising copy recognizes that although ". . . Holmes knew nothing of . . . modern-day technology when he wrote about a vehicle with no 'weak link' among its components," he did realize the absurdity of attempting to achieve "the perfect engineering feat."

In Holmes' poem, which starts on p. 35, the Deacon decides that he will build an indestructible shay, with every part as strong as the rest, so that it will not break down. However, what the Deacon fails to take into account is that everything has a lifetime, and if indeed a shay could be built with "every part as strong as the rest," then every part would "wear out" at the same time and whoever inherited the shay from the Deacon, who himself would pass away before his creation, would be taken by surprise one day. While "The Deacon's Masterpiece" is interesting in recognizing that breaking down is the wearing out of one part, the weakest link, it is not technologically realistic in suggesting that all parts could have exactly the same lifetime. That premise is contrary to the reality that we can only know that this or that part will last for *approximately* this or that many years, just as we can only state the probability that any one paper clip will break after so many bendings. The exact lifetime of a part, a machine, or a structure is known only after it has broken.

Just as we are expected to know our own limitations, so should we know those of the inanimate world. Even the pyramids in the land of the Sphinx, whose riddle reminds us that we all must crawl before we walk and that we will not walk forever, have been eroded by the sand and the wind. Nothing on this earth is inviolate on the scale of geological time, and nothing we create will last at full strength forever. Steel corrodes and diamonds can be split. Even nuclear waste has a half-life.

Engineering deals with lifetimes, both human and otherwise. If not fatigue or fracture, then corrosion or erosion; if not war or vandalism, then taste or fashion claim not only the body but the very souls of once-new machines. Some lifetimes are set by the intended use of an engineering structure. As such an offshore oil platform may be designed to last for only the twenty or thirty years that it will take to extract the oil from the rock beneath the sea. It is less easy to say when the job of a bridge will be completed, yet engineers will have to have some clear idea of a bridge's lifetime if only to specify when some major parts will have to be inspected, serviced, or replaced. Buildings have uses that are subject to the whims of business fashion, and thus today's modern skyscraper may be unrentable in fifty years. Monumental architecture such as museums and government buildings, on the other hand, should suggest a permanence that makes engineers think in terms of centuries. A cathedral, a millenium.

The lifetime of a structure is no mere anthropomorphic metaphor, for how long a piece of engineering must last can be one of the most important considerations in its design. We have seen how the constant on and off action of a child's toy or a light bulb can cause irreparable damage, and so it is with large engineering structures. The ceaseless action of the sea on an offshore oil platform subjects its welded joints to the very same back and forth forces that cause a paper clip or a piece of plastic to crack after so many flexures. The bounce of a bridge under traffic and the sway of a skyscraper in the wind can also cause the growth of cracks in or the exhaustion of strength of steel cables and concrete beams, and one of the most important calculations of the modern engineer is the one that predicts how long it will take before cracks or the simple degradation of its materials threaten the structure's life. Sometimes we learn more from experience than calculations, however.

Years after my son had outgrown Speak & Spell, and within months of his disaffection with the video games he once wanted so much, he began to ask for toys that required no batteries. First he wanted a BB gun, which his mother and I were reluctant to give him, and then he wanted a slingshot. This almost biblical weapon seemed somehow a less violent toy and evoked visions of a Norman Rockwell painting, in which a boy-being-a-boy conceals his homemade slingshot from the neighbor looking out a broken window. It is almost as innocent a piece of Americana as the baseball hit too far, and no one would want to ban slingshots or boys.

I was a bit surprised, however, to learn that my son wanted to *buy* a slingshot ready-made, and I was even more surprised to learn that his source would not be the Sears Catalog, which might have fit in with the Norman Rockwell image, but one of the catalogs of several discount stores that seem to have captured the imagination of boys in this age of high-tech toys. What my son had in mind for a slingshot was a mass-produced, metal-framed object that was as far from my idea of a slingshot as an artificial Christmas tree is from a fir.

Stephen was incredulous as I took him into the woods behind our house looking for the proper fork with which to make what I promised him would be a *real* slingshot. We collected a few pieces of trees that had fallen in a recent wind storm, and we took them up to our deck to assemble what I had promised. Unfortunately, I had forgotten how easily pine and dry cottonwood break, and my first attempts to wrap a rubber band around the sloping arms of the benign weapon I was making met with structural failure. We finally were able to find pieces strong enough to withstand the manipulation required for their transformation into slingshots, but their range was severely limited by the fact that they would break if pulled back too far.

My son was clearly disappointed in my inability to make him a slingshot, and I feared that he had run away disillusioned with me when he disappeared for an hour or so after dinner that evening. But he returned with the wyes of tree branches stronger and more supple than any I found behind our house. We were able to wrap our fattest rubber bands around these pieces of wood without breaking them, and they withstood as much pull as we were able or willing to supply. Unfortunately, they still did not do as slingshots, for the rubber bands kept slipping down the inclines of the Y and the bands were difficult to hold without the stones we were using for ammunition slipping through them or going awry.

After almost a week of frustration trying to find the right branch-and-rubber band combination that would produce a satisfactory slingshot that would not break down, I all but promised I would buy one if we could not make a top-notch shooter out of the scraps of wood scattered about our basement. Stephen was patient if incredulous as I sorted through odd pieces of plywood and selected one for him to stand upon while I sawed out of it the shape of the body of a slingshot. He was less patient when I drilled holes to receive a rubber band, and I acceded to his impatience in not sanding the plywood or rounding the edges before giving the device the test of shooting. I surprised him by producing some large red rubber bands my wife uses for her manuscripts, and he began to think he might have a real slingshot when I threaded the ends of a rubber band through the holes in the plywood Y. With the assembly completed I demonstrated how far a little pebble could be shot, but I had to admit, at least to myself, that it was very difficult to keep the pebble balanced on the slender rubber band. My son was politely appreciative of what I had made for him, but he was properly not ecstatic. The pebbles he tried to shoot dropped in weak arcs before his target, and he knew that his slingshot would be no match for the one his friend had bought through the catalog.

In my mind I admitted that the homemade slingshot was not well designed, and in a desperate attempt to save face with my son I decided to add a second rubber band and a large pocket to improve not only the range but also the accuracy of the toy. These proved to be tremendous improvements, and with them the sling-shot seemed almost unlimited in range and very comfortable to use. Now we had a slingshot of enormous potential, and my son was ready to give it the acid test. We spent an entire weekend practicing our aim at a beer bottle a good thirty yards away. The first hit was an historic event that pinged off the glass and the second a show of power that drilled a hole clear through the green glass and left the bottle standing on only a prayer. As we got better at controlling the pebbles issuing from our homemade slingshot we changed from bottles to cans for our targets and hit them more and more.

With all our shooting, the rubber bands began to break from fatigue. This did not bother my son, and he seemed to accept it as something to be expected in a slingshot, for it was just another toy and not a deacon's masterpiece. As rubber bands broke, we replaced them. What proved to be more annoying was the slipping of the rubber band over the top of the slingshot's arm, for we had provided no means of securing the band from doing so. In time, however, we came to wrap the broken rubber bands around the top of the arms to keep the functioning ones in place. This worked wonderfully, and the satisfaction of using broken parts to produce an improved slingshot was especially appealing to my son. He came to believe that his slingshot could outperform any offered in the catalogs, and the joy of producing it ourselves from scrap wood and rubber bands gave him a special pleasure. And all the breaking pieces of wood, slipping rubber bands, and less-than-perfect functioning gave him a lesson in structural engineering more lasting than any textbook's—or any fanciful poem's. He learned to make things that work by steadily improving upon things that did not work. He learned to learn from mistakes. My son, at eleven, had absorbed one of the principal lessons of engineering, and he had learned also the frustrations and the joys of being an engineer.

APPENDIX

THE DEACON'S MASTERPIECE
Or, the Wonderful "One-Hoss Shay"
A Logical Story
By Oliver Wendell Holmes

Have you heard of the wonderful one-hoss shay,
That was built in such a logical way
It ran a hundred years to a day,
And then, of a sudden, it—ah, but stay,
I'll tell you what happened without delay,
Scaring the parson into fits,
Frightening people out of their wits—
Have you ever heard of that, I say?

Seventeen hundred and fifty-five.
Georgius Secundus *was then alive,—*
Snuffy old drone from the German hive.
That was the year when Lisbon-town
Saw the earth open and gulp her down,
And Braddock's army was done so brown,
Left without a scalp to its crown.
It was on the terrible Earthquake-day
That the Deacon finished the one-hoss shay.

Now in the building of chaises, I tell you what,
There is always, somewhere, *a weakest spot,—*
In hub, tire, felloe, in spring or thill,
In panel, or crossbar, or floor, or sill,
In screw, bolt, thoroughbrace,—lurking still,
Find it somewhere you must and will,—
Above or below, or within or without,—
And that's the reason, beyond a doubt,
That a chaise breaks down, *but doesn't* wear out.

But the Deacon swore (as deacons do,
With an "I dew vum," or an "I tell yeou")
He would build one shay to beat the taown
'N' the keounty 'n' all the kentry raoun';
It should be so built that it could n' *break daown:*
"Fur," said the Deacon, " 't 's mighty plain
Thut the weakes' place mus' stan' the strain;
'N' the way t' fix it, uz I maintain, Is only jest
T' make that place uz strong us the rest."

So the Deacon inquired of the village folk
Where he could find the strongest oak,
That couldn't be split nor bent nor broke,—
That was for spokes and floor and sills;
He sent for lancewood to make the thills;
The crossbars were ash, from the straightest trees,
The panels of white-wood, that cuts like cheese,
But lasts like iron for things like these;
The hubs of logs from the "Settler's ellum,"—
Last of its timber,—they couldn't sell 'em,
Never an axe had seen their chips,
And the wedges flew from between their lips,
Their blunt ends frizzled like celery-tips;
Step and prop-iron, bolt and screw,

Spring, tire, axle, and linchpin too,
Steel of the finest, bright and blue;
Thoroughbrace bison-skin, thick and wide;
Boot, top, dasher, from tough old hide
Found in the pit when the tanner died.
That was the way he "put her through."
"There!" said the Deacon, "naow she'll dew!"

Do! I tell you, I rather guess
She was a wonder, and nothing less!
Colts grew horses, beards turned gray,
Deacon and deaconess dropped away,
Children and grandchildren—where were they?
But there stood the stout old one-hoss shay
As fresh as on Lisbon-earthquake-day!

EIGHTEEN HUNDRED; it came and found
The Deacon's masterpiece strong and sound.
Eighteen hundred increased by ten;—
"Hahnsum kerridge" they called it then.
Eighteen hundred and twenty came;—
Running as usual; much the same.
Thirty and forty at last arrive,
And then come fifty, and FIFTY-FIVE.

Little of all we value here
Wakes on the morn of its hundredth year
Without both feeling and looking queer.
In fact, there's nothing that keeps its youth,
So far as I know, but a tree and truth.
(This is a moral that runs at large;
Take it.—You're welcome.—No extra charge.)

FIRST OF NOVEMBER,—the earthquake-day,—
There are traces of age in the one-hoss shay,

A general flavor of mild decay,
But nothing local, as one may say.
There couldn't be,—for the Deacon's art
Had made it so like in every part
That there wasn't a chance for one to start.
For the wheels were just as strong as the thills,
And the floor was just as strong as the sills,
And the panels just as strong as the floor,
And the whipple-tree neither less nor more,
And the back crossbar as strong as the fore,
And spring and axle and hub encore.
And yet, as a whole, *it is past a doubt*
In another hour it will be worn out!

First of November, 'Fifty-five!
This morning the parson takes a drive.
Now, small boys, get out of the way!
Here comes the wonderful one-hoss shay,
Drawn by a rat-tailed, ewe-necked bay.
"Huddup!" said the parson.—Off went they.
The parson was working his Sunday's text,—
Had got to fifthly, *and stopped perplexed*
At what the—Moses—was coming next.
All at once the horse stood still,
'Close by the meet'n'-house on the hill.
First a shiver, and then a thrill,
Then something decidedly like a spill,—
And the parson was sitting upon a rock,
At half past nine by the meet'n'-house clock,—
Just the hour of the Earthquake shock!
What do you think the parson found,
When he got up and stared around?
The poor old chaise in a heap or mound,
As if it had been to the mill and ground!

You see, of course, if you're not a dunce,
How it went to pieces all at once,—
All at once, and nothing first,—
Just as bubbles do when they burst.

End of the wonderful one-hoss shay.
Logic is logic. That's all I say.

4
ENGINEERING
AS HYPOTHESIS

Every issue of *The Structural Engineer,* the official journal of the British Institution of Structural Engineers, carries prominently displayed in a box on its contents page this definition of its subject:

> Structural engineering is the science and art of designing and making, with economy and elegance, buildings, bridges, frameworks, and other similar structures so that they can safely resist the forces to which they may be subjected

Since some engineers deny that engineering is either science *or* art, it is encouraging to see this somewhat official declaration that it is both. And indeed it is, for the *conception* of a design for a new structure can involve as much a leap of the imagination and as much a synthesis of experience and knowledge as any artist is required to bring to his canvas or paper. And once that design is articulated by the engineer as artist, it must be analyzed by the engineer as scientist in as rigorous an application of the scientific method as any scientist must make.

The British Institution's definition of structural engineering crowds into the same box the ideas of economy and elegance, for responsible engineering wastes neither physical nor mental resources. Economic constraints are often imposed by the demands of the marketplace, but the requirement for elegance is often self-imposed by the best in the profession in much the same way that artists and scientists alike see elegance in the sparest canvases and the most compact theories—or in the axiom of minimalist aesthetics and design, "less is more."

Finally, the definition concludes with the idea of safety, an objective that is ultimately more important than either the economic or aesthetic ones, for the loss of a single life due to structural collapse can turn the most economically promising structure into the most costly and can make the most beautiful one ugly. The structural engineers' definition comes to an end with the idea that structures are safe if they can "resist the forces to which they may be subjected," but it is symbolic of the virtually endless list of forces to which a structure might in fact be subjected that there is no period at the end of what seems otherwise to be a complete sentence.

The idea of resisting forces is a simple one, but putting it into practice can be rather tricky. On the one hand, the resistance of materials that form the building blocks of the engineer is not ever precisely known, for there is always the danger of the weak link in an otherwise sound chain. On the other hand, predicting the forces to which a structure may be subjected at any time in the future can be as difficult and as unsure as predicting the weather that may be responsible for some of those forces. Hence structural engineering must often deal in probabilities and combinations of probabilities. A safe structure will be one whose weakest link is never overloaded by the greatest force to which the structure is subjected. In order to make such ideas precise enough to be able to order the right amount of steel or concrete for a structure, the engineer must be prepared to posit causes and anticipate as much as possible what forces besides gravity will be acting on his proposed structure, which is like a little man-made solar system created to be set among other man-made systems within the given universe.

The object of a science may be said to be to construct theories about the behavior of whatever it is that the science studies. Observation and experience, inspiration and serendipity, genius and just good guesses—by their presence and absence, in pinches and dashes—all can provide the recipe for a scientific theory. As with all recipes, in which the cook is always the invisible ingredient, the individuality of the scientist provides the inexpressible human flavor. This aspect of science, the concoction of theories, has no universal method. But once a theory has evolved, perhaps from a half-baked idea to a precise and unambiguous statement of the scientist's entry in the great universal cook-off, the scientific method may be used to judge the success or failure of a given theory or the relative merits of competing theories. Theories entered in the scientific cooking contest are known as hypotheses, and the process of judging is known as the testing of hypotheses.

A scientific hypothesis is tested by comparing its conclusions with the reality of the world as it is. Yet, no matter how many examples of agreement one may collect, they do not prove the truth of the hypothesis, for it may be argued that one has not tested it in the single case where the theory may fail to agree with reality. On the other hand, just one instance of disagreement between the hypothesis and reality is sufficient to make the hypothesis incontrovertibly false. That honeybees always build their hives with hexagonal cells is a hypothesis that has accumulated so much verification that it is hardly called a hypothesis anymore. It is assumed to be a *fact*. But let some young apiarist discover his bees making octagonal cells, and not only would the hypothesis that bees always use the hexagon be forever smashed, but there would also be quite a bit of excitement among the world of honeybee experts. That the sun rises each morning may also be considered a hypothesis, and our experience that indeed this happens day in and day out serves to confirm—but not prove—the hypothesis. Yet all it would take would be a single "morning" without a sunrise to make the contention that the sun rises every morning categorically false. While it may be beyond our comprehension that this could ever be the case, it nevertheless remains true that our belief that the sun will rise tomorrow is basically a matter of faith rather than of rigorously established fact.

Just as winning the blue ribbon at the county fair is no guarantee that a cake will be everybody's favorite at next year's fair, so the present success of a theory, even one so mundane as the diurnal appearance of the sun, is no assurance of its continued success. Hence, Newton *seemed* to have had the last word on theories of planetary motion for over two centuries, in spite of some intractable phenomena, until Einstein proposed his own more general theory. While it may be the conventional wisdom that Einstein had finally got the workings of the universe right, the history of science strongly suggests that this is presumptuous and highly unlikely. What may someday supersede Einstein's hypothesis is any genius' good guess. In the meantime, not only the theory of relativity but also Newton's laws, with all their known limitations, serve us rather well in navigating through space and in constructing bridges and dams on earth. It is one of the marvels of the practice of engineering and of science that one can accomplish so much with so few and such admittedly approximate theories as Newtonian mechanics.

Engineering design shares certain characteristics with the positing of scientific theories, but instead of hypothesizing about the behavior of a given universe, whether of atoms, honeybees, or planets, engineers hypothesize about assemblages of concrete and steel that they arrange into a world of their own making. Thus each new building or bridge may be considered to be a hypothesis in its own right. In particular, one hypothesis of a structural engineer might be that so and so bridge across such and such river under these and those conditions of traffic and maintenance will stand for so many years without collapsing. Now if such a bridge were built and were to carry traffic year after year without trouble, the hypothesis would be confirmed time and time again—but it will never be *proven* until the so many years under the original plan had elapsed. But should the bridge collapse suddenly under no extraordinary conditions before those so many years were up, there would be no doubt in anyone's mind that the original hypothesis was incontrovertibly wrong.

The process of engineering design may be considered a succession of hypotheses that such and such an arrangement of parts will perform a desired function *without fail.* As each hypothetical arrangement of parts is sketched either literally or figuratively on the calculation pad or computer screen, the candidate structure must be checked by analysis. The analysis consists of a series of questions about the behavior of the parts under the imagined conditions of use after construction. These questions may be easily answered for designs that are not particularly innovative, but a computer may be required to perform all the calculations needed to analyze a bold new design. If any of the parts fails the test of analysis, then the design itself may be said to be a failure. A design can be altered by strengthening the weak link and then analyzing the new design. The process continues until the designer can imagine no possible way in which the structure can fail under the anticipated use. Of course, if the designer makes an error in calculation or overlooks some possibility of failure or does not program the computer to ask the right question, then the hypothesis will erroneously be thought to have been verified when in fact it should have been disproved. Absolute certainty about the fail-proofness of a design can never be attained, for we can never be certain that we have been exhaustive in asking questions about its future.

The fundamental feature of all engineering hypotheses is that they state, implicitly if not explicitly, that a *designed* structure will not fail if it is used as intended. Engineering failures may then be viewed as disproved hypotheses. Thus, the failure of the Hyatt Regency elevated walkways disproved the hypothesis that those skywalks could support the number of people on them at the time of the collapse; the failure of the Tacoma Narrows Bridge disproved the hypothesis that the suspension span could carry morning traffic in a forty-two-mile-per-hour crosswind; and the failure of the Teton Dam disproved the hypothesis that it could hold back river water for irrigation. On the other hand, the past success of an engineering structure confirms the hypothesis of its function only to the same extent that the historical rising of the sun each morning has reassured us of a predictable future. The structural soundness of the Brooklyn Bridge only proves to us that it has stood for over one hundred years; that it will be standing tomorrow is a matter of probability, albeit high probability, rather than one of certainty.

Such realizations need be no more disturbing than those associated with the probabilities of disease. Indeed, if we find ourselves having led healthy lives for so many years, we all know deep in our hearts that that is no guarantee that we will not be hospitalized, or worse, tomorrow. We have seen members of our families and friends struck suddenly with cancer or back trouble or a heart attack or the bite of an insect. Or we have known someone in his or her prime struck down by a car or taken as the victim of an airplane crash or a freak accident or even a structural failure. If we were immobilized by the fear of such a fate, we might have to be institutionalized, for we would cease to be able to function in the world as it is. Lightning strikes, and some time it might strike us. We must accept this as a cost of the pleasures of life, and we have no choice. We risk remote dangers every day for the benefit of the pleasures of the day.

One of the most elementary and most common of structural forms is what engineers call a *beam*. The essence of the idea of a beam is that it spans some space and resists bending or deflection by forces acting transverse to its long dimension. Some familiar examples will make this somewhat abstract definition concrete. Houses have floor beams, and if we examine them in an unfinished basement we see that they generally reach from wall to wall, sometimes supported by intermediate columns when the walls are especially far apart. The beams support their own weight and the weight of the floor above along with that of the furniture and people that rest and move about on it, which tends to cause the beams to dip, usually imperceptibly, toward the ground. Since ordinary houses have ordinary dimensions, a modern house builder thinks very little about the size or strength of floor beams, for he just builds each new house pretty much as he has built all his previous successful houses.

But what might we have to consider if we were stranded on a desert island with only the lumber for a house and not the plans or experience of a builder? First we might notice as we move the lumber about that a long two-by-ten bends rather easily in one direction but not in the other. The phenomenon is exaggerated in a yardstick that, being smaller and weaker than the piece of lumber, responds more easily to the action of our hands. Thus if we experiment with the yardstick we notice that when it is flat it can easily be depressed, and we can easily imagine that the stick will sag ever so slightly in the middle if we support it on our fingers at the one- and thirty-five-inch marks. An engineer would say, "The simply supported beam is uniformly loaded by only its own weight." While we would not construct our house out of such pliant yardsticks, we can imagine that our two-by-tens will sag in a similar fashion, especially when more than their own weight rests upon them, and should we build our floor on the beams' flat sides, we should expect not only a noticeable sag but also an unsettling springiness as we walked about the living room. On the other hand, if we were to support the yardstick near its ends with its advertisement vertical, as if on a billboard, we would notice no appreciable sag. An engineer would say, "The beam is much more resistant to bending under its own and other weight in this orientation," and we can generalize this experience and experiment to say that floor beams should be installed so that they present their deep and not their flat sides to the weight of the floor.

The action of a heavy piece of furniture or people on the floor can be examined by holding the yardstick at its ends and pushing with our knee against the middle. We should not only see the much greater bending when our knee pushes against the flat side; we should also notice that the yardstick will tend to twist out of shape if we do not take care to hold it straight when our knee pushes against the sharper side. The engineer calls this kind of distortion an "instability" or "buckling," and we can feel that the phenomenon appears in the yardstick all of a sudden as we increase the force of our knee. To be sure that our house's floor beams will not exhibit this tricky action, we might brace them at one or two places along their span to offer resistance to undesirable motion of the beam. Now whether or not we anticipated all these actions in our first attempt at house building on a desert island might depend on whether or not the sagging of the lumber struck us as significant, whether or not we happened to play with a yardstick as we considered the construction task before us, and whether or not we tried to lay down our floor beams in the less convenient way. For if we had no man Friday, it might certainly be easier to lay the beams flat than to attempt to balance a bunch of them in the top-heavy position while we laid a floor over them. That might take the patience and luck of a builder of a house of cards.

Something so common as the construction of a floor may thus be viewed as the statement of a hypothesis, albeit one reached possibly by trial and error and never made quite explicit. When we install our beams in a certain way we say implicitly that they will support the floor without excessive sagging, without snapping out of place, and without breaking. If we did not appreciate the different action of a beam laid flat and a beam balanced deep, if we did not expect a beam to snap out of place when we moved in a big flat boulder for a coffee table, or if we did not expect to exceed a beam's breaking point when we jumped up and down doing our morning exercises, then we might find our hypothesis disproven in a funnel of rubble in the basement. On the other hand, if our house does stand, that does not necessarily mean that we had, by luck or by design, hit on the optimal way to lay floor beams. We might have laid them flat over a relatively short span and not imagined anything undesirable or unavoidable in the slight sag and bounce we noticed as we walked across the room. Or we might not miss the bracing because we abhorred large coffee tables. Or we might do our exercise out of doors and thus never test the floor's limits of endurance. Thus we might never test the hypothesis of our floor construction with a critical experiment.

Should another person then be shipwrecked on our island with another load of lumber, we would naturally have our experience in house building to share with him. But his lumber might be of longer lengths or of more slender dimensions, and he might want to build a larger house than ours. Had we had any structural failures, we would have tested certain hypotheses and found them to be wanting. Thus we could tell the neophyte what *not* to do. However, if our original house were still standing, our neighbor might just copy our design, scaling it up and employing his longer boards in his more ambitious structure. And the house might stand if the boards were not too much longer or too much weaker or if the new resident were not so heavy as we and not so much disposed to furnish his house with boulders as with bamboo tables and chairs. But should he someday decide to redecorate, and should the two of us drop the boulder sofa in the middle of the floor, we might disprove the hypothesis that his house is solid as a rock.

Now a house can fail in ways other than having its floor collapse, and we can imagine that walls and roofs can have their own weaknesses. These can generally be avoided by copying successful designs, but as with our island occupants, deviating from an exact copy can be disastrous. And deviations inevitably occur, through carelessness, greed, and well-intentioned new hypotheses to build bigger houses more quickly and with less lumber. But the builder who is overconfident in his new departures or who does not brace his frame adequately may find it flat on the ground after a midnight storm. And a bad winter can so overload roofs with snow that their collapses become endemic. The winter of 1979 saw so much snow accumulate on barn roofs around Chicago that a number of them that had stood through many a more routine winter collapsed.

If building a house or barn is fraught with such dangers, what must be the task of the engineer charged with building a record-length bridge, which may be thought of as just one great big beam, or a record-high skyscraper, which may be thought of as a tall, tall beam rooted in the ground. In these cases there is nothing to copy, and there may be little relevance to experimenting with the likes of yardsticks or even of twigs. Even should the engineer have had the opportunity to experience similar but less ambitious structures than the one he is called upon to design, there always remains the question as to how validly he can extrapolate his experience and how far he can go beyond the last confirmed hypothesis. It is here that engineering comes to resemble science in another way, for engineers find it necessary to study beams and other elements of construction as if they were the natural stuff of science. Indeed, the discipline of taking the universe of structural elements as one's field of study is known as engineering science, and it has a long and separate history from that of pure science.

Galileo was working in the spirit of modern engineering scientists when he considered the resistance of solids to fracture in the second day of his *Dialogues Concerning Two New Sciences.* Among the matters Salviati discusses with Sagredo and Simplicio is the strength of what is known today as a cantilever beam. This is a beam held only at one end while supporting weight or resisting motion all along its free length. Trees and skyscrapers act as cantilever beams when they resist the pressure of the wind tending to bend and topple them. Our outstretched arms, holding biscuits just out of our dog's reach, are also acting as cantilever beams, as are flagpoles and balconies. Galileo's cantilever resembles a piece of lumber embedded at one end in a section of a masonry wall and supporting a great boulder from a hook at the beam's extremity. The seventeenth-century drawing is often reproduced, and its embellishments of vegetation and shadows put to shame many of the bare illustrations in today's engineering textbooks.

But if the classic sketch of Galileo's cantilever is ornate, his analysis of the beam's strength is spare and no-nonsense. He correctly observed, no doubt after actually breaking some cantilever beams, at least in his mind, much as our shipwrecked house builder might have broken some yardsticks, that as the weight is increased the beam cracks and breaks at the juncture with the wall. But since Galileo was apparently the first to attack this problem in a rational way, it is perhaps understandable that he made some mistaken assumptions about how the breaking force is actually distributed across the depth of the beam. (That would not be known correctly for another seventy-five years, when in 1713 a Frenchman named Parent would publish two memoirs on the bending of beams.) Nevertheless, even with his mistaken notions about the way the beam resists breaking, Galileo came to the correct conclusion that the beam's strength is proportional to the square of the depth of the beam. This is consistent with our experience that it is much easier to break a piece of lumber by bending it through its small rather than its large dimension, and a one-by-ten does indeed appear to have a hundred times as much resistance to bending through its ten- as across its one-inch depth. Thus Galileo's result tells us what we already know about a yardstick held in our fist at one end only: the stick will resist being bent more one way than another. Trees, those organic vertical cantilevers that are subject to bending by the wind, conveniently have near-circular trunks that give equal resistance to the wind no matter what its direction. Skyscrapers, on the other hand, are generally not circular in plan, for structural considerations such as wind resistance are seldom as dominant as architectural or functional factors in determining the shape of a tall building.

Galileo's analysis of the cantilever beam illustrates an extremely important point for understanding how structural accidents can occur: he arrived at what is basically the right qualitative answer to the question he posed himself about the strength of the beam, but his answer was not absolutely correct in a quantitative way. He got the right qualitative answer for the wrong quantitative reason. Thus Galileo could correctly have advised any builders how to orient their beams for the best results, but should he have been asked to predict the absolute minimum-sized beam required to support a certain weight so many feet out from a wall, the answer he calculated from his formula might have been too weak by a factor of three. We shall come back to this kind of error when we deal with the concept of factors of safety, but the important point here is that *apparently* right answers can be gotten through wrong reasoning. As a very simple illustration of this phenomenon, consider the theory that the product of a number with itself is the same as the sum of the number and itself. This theory works perfectly when the number is 2, for $2 \times 2 = 4$ gives the same result as the calculation $2 + 2 = 4$.

If one were so inclined to believe this theory that one did not test the hypothesis beyond the single case of the number 2, one might believe one were correctly calculating the squares of all numbers by merely doubling them. Consider further that a number squared represents the amount of load applied to a beam. One might even get away with such an error as long as one had no need for great accuracy and had no need to square numbers other than those close to 2. However, if one day the square of 20 were needed, the erroneous method would give only 40 instead of 400, and the order-of-magnitude mistake might finally be uncovered because the beam would break when the theory predicted it should not have. Such a failure would indeed contribute more to uncovering error than all the successful "verifications" of the incorrect hypothesis.

It is easy now for sophomore engineering students taking their first course in strength of materials to see the error in Galileo's analysis, but that is not to say that the error should have been obvious to him or his contemporaries. Hindsight is always 20/20, but most of us have at some time experienced myopia when we have had to stand back and criticize our own work. Who has not been amazed at the typographical error missed in countless proof-readings? Who has not been frustrated by the arithmetical error that stands through numerous attempts to balance a checkbook?

Engineers today, like Galileo three and a half centuries ago, are not superhuman. They make mistakes in their assumptions, in their calculations, in their conclusions. That they make mistakes is forgivable; that they catch them is imperative. Thus it is the essence of modern engineering not only to be able to check one's own work, but also to have one's work checked and to be able to check the work of others. In order for this to be done, the work must follow certain conventions, conform to certain standards, and be an understandable piece of technical communication. Since design is a leap of the imagination, it is engineering analysis that must be the lingua franca of the profession and the engineering-scientific method that must be the arbiter of different conclusions drawn from analysis. And there will be differences, for as problems come to involve more complex parts than cantilever or even simply supported beams, the interrelations of those various parts not only in reality but also in the abstract of analysis become less and less intuitive. It is not so easy to get a feel for a mammoth structure like a jumbo jet or a suspension bridge by flexing a paint store yardstick in one's hands. And the hypothesis that a structure will fly safely through wind and rain can be worth millions of dollars and hundreds of lives.

5
SUCCESS IS
FORESEEING FAILURE

All the successes of engineering as far back in history as the pyramids and as far into the future as the wildest conceptions of mile-high skyscrapers may be imagined to have begun with a wish to achieve something *without failure,* where "without failure" to the engineer means not only to stand without falling down but also to endure with what might be called "structural soundness." Unsound structures—those that are eaten away by rapid corrosion, those that have repeated service breakdowns under ordinary conditions, those that suffer from fatigue cracking after not so many years of use—may be thought to have been failures as surely as if they had collapsed during construction. And no matter how ingenious or attractive his conception may appear in his imagination or on paper, if a designer overlooks just one way in which his structure may fail, all may be for naught.

The earliest engineering structures may have been designed by trial and error, and it may be argued that the Egyptian pyramids were built using that method. Although the pyramids' construction process, with all its staggering numbers and sizes of blocks of stone and crews of workers, may never be known with certainty, it is not too hard to imagine why the shapes are what they are. The pyramid shape is an extremely stable one, perhaps suggested by the shape taken by a pile of sand at the bottom of an hourglass. But it is a timeless shape, one that resembles mountains, and one that in its monolithic appearance when viewed from afar looks as if it can resist being toppled over in even the fiercest sandstorm. And even though the pyramid is not a regular tetrahedron and thus not one of the Platonic solids, it is certainly an appealing if not a mystical shape. The Egyptian pyramids have more or less square bases and rise in square-cornered tiers, which is a much more natural way to assemble squarish blocks of stone than in the triangular pattern a tetrahedron would demand. Yet even with all of these more or less natural decisions made willy-nilly, the decision of exactly how to pile the stones and exactly how steep to make the sides of the pyramid rise is a crucial one that is not so naturally arrived at.

A pile of sand will assume a natural conical angle as the sand is dripped from the fist, but that angle can vary with the kind of sand and with the conditions under which the pile rises. If the pile rises on a patch of desert that is itself but a greater pile of sand, the added weight can lead to little avalanches, as beach sand acts when we try to pile it too high. The earliest pyramids are believed to have evolved from mastabas, which were rather low, rectangular tombs enclosed with sloping walls of brick. About 2700 B.C. Imhotep, the first builder known by name, apparently was charged with constructing a tomb for the Egyptian Pharaoh Zoser of the Third Dynasty, and Imhotep chose to elaborate on the mastabas of the First Dynasty. He first faced a conventional mastaba design with stone, then piled stone upon its top in what might be called a stepped pyramid. Pharaoh Zoser's pyramid thus grew in stages, with Imhotep perhaps gaining confidence to add more and more stone as each succeeding course did not fail.

Once Imhotep succeeded in raising a stepped pyramid, others could copy it with confidence against failure. But, apparently not satisfied with a pyramid with great steps, subsequent designers elaborated on Imhotep's success and the stepped profile evolved into one with the steps filled in to give the familiar straight edge and flat sides now commonly associated with the pyramids of

Egypt. The success of the Meidum pyramid, whose faces rose more steeply than those of its predecessors, gave still later designers something else to better. They attempted to do this in the pyramid at Dahshur, whose sides began to rise at the previously untried angle of 54°. However, something appears to have happened during construction to change the original plan, for approximately halfway up, the Dahshur pyramid suddenly changes not only in the kind of stone used but also in the slope of its sides. From 54° the walls drop to a 43° inclination—hence the structure's descriptive name, the Bent Pyramid. One theory holds that the break in the Bent Pyramid resulted from a structural failure attributed to construction at the theretofore untried steeper angle, and that the designer's original aspirations had to be lowered. What has been interpreted as great masses of debris about the base of the Bent Pyramid lends credence to this early example of structural design being pushed by hubris to the limits of failure. Subsequent pyramid builders, who built higher but dared not build steeper, seemed to be content with humbler successes.

Egyptian pyramid builders are not unique in their encounters with the limits of structures and their desire to do what had not been done before. Medieval cathedrals were certainly much more complex structurally than the pyramids, yet there is still considerable evidence that the cathedrals evolved through a process of experimentation and trial and error not unlike that of the Egyptian megaliths. Even the layman Henry Adams, in his avuncular Baedeker to the cathedrals of France, had to remark on how the architects of churches erected only forty or fifty miles apart around Paris in the late twelfth and early thirteenth centuries must have watched and been influenced "almost from day to day" by each other's experiments. One builder's structural and aesthetic successes and failures were challenges and lessons to the others.

In the year 1284 the cathedral at Beauvais suffered a major collapse, and this incident has been regarded as a turning point in the development of Gothic structures. Thenceforth architects are generally assumed to have been more conservative in their structural adventures, although thoughtful critics see a resurgence of structural innovation and aspiration in the fourteenth century. Robert Mark, who has used modern engineering models to analyze the forces acting on Gothic cathedrals, sees significant new achievements in height and slenderness exhibited in the nave of the Palma, Majorca, cathedral, for example, though he acknowledges that the accomplishments are by no means as major as those that culminated in Beauvais.

But pyramids and cathedrals may be said to belong to the pre-rational age of structural engineering, for there was apparently much more reliance on physical experiment and mid-construction correction than on any predetermined and inviolate set of plans for the final version of the structure. The flying buttresses of the cathedrals seem certainly to have been added and elaborated upon in response to undesirable cracking in the medieval masonry. Even the addition of apparently decorative pinnacles seems to have been in response to a functional need for more weight to keep further cracks from opening up under the great forces of the wind to which a massive cathedral was subjected as it rose exposed from among all the ground-hugging buildings of a medieval town.

The great structures of the nineteenth and twentieth centuries are the iron and concrete bridges and skyscrapers that have a slenderness and structural daring perhaps undreamed of by the builders in stone. The tenacity of steel has added a tensile dimension to structures that brings a release from the dominance of compression as a stabilizing force almost as if from the pull of gravity itself, but the memory of countless iron bridge failures during the nineteenth century keeps modern engineers from getting swell-headed over how fast they can build longer and taller structures even today. Indeed, the chronic ailments of iron bridges that evolved with the expansion of the railroads is to this day one of the most discussed and most chronicled chapters in the history of structural engineering.

Perhaps this is so because of the differing symbolic nature of bridges compared with monumental architecture such as pyramids and cathedrals. The latter two were erected as tributes to earthly and heavenly rulers, but bridges are principally functional structures. When they do assume symbolic proportions, it is almost always as an afterthought, as when the Brooklyn Bridge took its place in art and literature only after it had become firmly established as political link across the East River. If anything, the modern bridge is a tribute to man's technological achievements, and no other structure shows off its structural bones or flexes its structural muscles the way a bridge does.

The nineteenth-century expansion of the railroad brought new challenges to man the builder. Pyramids were by and large great piles of stone, and the only spaces the stone was required to span were the long but narrow labyrinthine corridors leading to burial chambers no more ambitious in scale than a bedroom for today's common man. Cathedrals, on the other hand, along with their great domed predecessors in Rome, strove not for height with mass and bulk but for height with delicacy and openness. Neither economics nor mass production was the first concern of the builders of the monumental architecture, but crass realities became principal ingredients in the evolution of bridges. That is not to say that there was less concern for an iron bridge to be successful than for a Gothic cathedral. Indeed, driven by economics, the railroad companies were perhaps even less inclined to risk structural failure than were the patrons of cathedrals, whose motivation lay beyond material profit or loss. Yet it might also be said that the railroad companies were still more daring, for they had to construct bridges under ever-new conditions, and merely copying what had stood the test of time was not always an option.

With the growth of the railroads came new demands on engineering structures. Railroad bridges had to withstand not only the sheer weight of the heavy locomotives and the rest of the rolling stock but also the dynamic action of the engine's reciprocating parts and the constantly changing position of the train on the bridge. Railroads brought together the machines of the mechanical engineer and the stationary structures of the civil engineer, and the demands of each provided stimulus for the evolution of the other—but not without mishap. As the tentacles of the railroads reached further and further out to provide service, there were more and more incentives to have heavier trains traveling at greater speeds over more rugged terrain. Any hill that did not have to be climbed or any valley that did not have to be entered was time and energy saved, and that meant money. But soon the strength required for a railroad bridge built to accommodate an early generation of iron horses was exceeded by later, heavier generations, and collapses occurred. Each defective bridge resulted in demands for excess strength in the next similar bridge built, and thus the railroad bridge evolved through the compensatory process of trial and error. As Ralph Waldo Emerson observed in his contemporaneous essay on compensation, "Every excess causes a defect; every defect an excess."

There was a technological awareness in the nineteenth century that recognized railroad trains and their infrastructure as a part of the changing environment. Thus William Wordsworth, the worshipper of nature, wrestled with the problem of a new culture intruding into the older, more pastoral one. The Industrial Revolution was changing the face of the English countryside, and in the poem, "On the Projected Kendal and Windermere Railway," he exhorts the reader to share his disdain for what Wordsworth views as a disfigurement of the Lake District:

> *Hear Ye that Whistle? As her long-linked Train*
> *Swept onwards, did the vision cross your view?*
> *Yes, ye were startled;—and in balance true,*

> *Weighing the mischief with the promised gain,*
> *Mountains, and Vales, and Floods, I call on you*
> *To share the passion of a just disdain.*

But disdain was tempered in another poem, entitled "Steamboats, Viaducts and Railways," in which Wordsworth acknowledged technology to be but another manifestation of a greater nature:

> *In spite of all that beauty may disown*
> *In your harsh features, Nature doth embrace*
> *Her lawful offspring in Man's art; and Time,*
> *Pleased with your triumphs o'er his brother Space,*
> *Accepts from your bold hands the proffered crown*
> *Of hope, and smiles on you with cheer sublime.*

Wordsworth's honest ambivalence between improved and unimproved Nature was the poet's response to the engineer's introduction of amendments to a given nature. Indeed, Wordsworth, in a later version of the last line, replaced the somewhat distant "smiles on you" with the more positive "welcomes you." His expression was as subject to improvement as the bridges of the nineteenth-century engineers. But the numerous bridge failures did not make it any easier for Wordsworth or any of his contemporaries to embrace the new technology. They were all too aware of the trying and erring going on not only in Britain but also in America.

In 1843 Nathaniel Hawthorne took *The Pilgrim's Progress,* John Bunyan's seventeenth-century religious allegory of the good man's pilgrimage through life, as a model for a tale set in the midst of the Industrial Revolution. In "The Celestial Railroad" Hawthorne's traveler is accompanied from the City of Destruction to the Celestial City by a Mr. Smooth-it-away, who points out how the new technology might improve the condition of man, which had escaped improvement for so long. But the traveler seems constantly distracted by the condition of the bridges over which he travels. One, though "of elegant construction," he imagined was "too slight . . . to sustain any considerable weight." Another "vibrated and heaved up and down in a very formidable manner; and, spite of Mr. Smooth-it-away's testimony to the solidity of its foundation," the traveler "should be loath to cross it in a crowded omnibus, especially if each passenger were encumbered with . . . heavy luggage." Structural failure is also on the fictional character's mind as the Hill Difficulty is approached:

> Through the very heart of this rocky mountain, a tunnel had been constructed of most admirable architecture, with a lofty arch and a spacious double track; so that, unless the earth and rocks should chance to crumble down, it will remain an eternal monument of the builder's skill and enterprise. It is a great though incidental advantage that the materials from the heart of the Hill Difficulty have been employed in filling up the Valley of Humiliation, thus obviating the necessity of descending into that disagreeable and unwholesome hollow.

It was a "wonderful improvement, indeed."

The ambivalence expressed by such writers as Wordsworth and Hawthorne was echoed in the popular press of the time. On the one hand, the fruits of the Industrial Revolution were plucked with anticipation and laid out graphically in all their succulence for the public to savor in such chronicles of the times as the *Illustrated London News* and *Harper's Weekly*. On the other hand, *Harper's* would report the accidents that occurred and *Punch* would satirize and parody the railroads. The railroads and their bridges had captured both the imagination and the fear of the public, just as airplanes would a century later. These technological advances were for Everyman and not just for kings and

God, and their promise of a smoother route to the Celestial City presented benefits that made the risk of accidents along the way a risk worth taking. For all of Hawthorne's ridicule, his first-person narrator does want to take a ride on the Celestial Railroad "to gratify a liberal curiosity."

As the *Oxford English Dictionary* attests, the word *engineer* designated one who contrives, designs, or invents more than a century before it came to mean also one who manages an engine. The latter meaning dates from 1839, when the railroad was emerging as the great metaphor of the Industrial Revolution, and it is not surprising that there came to be a deliberate confusion of the contriver and the driver of the vehicle. The engineers of steam engines and iron bridges were in the driver's seat. As these mechanical and structural pioneers were pushing the railroad further and further beyond the frontiers of technology, they were increasingly seen as controlling the speed and destination of the passengers on the Celestial Railroad. Although Hawthorne's alter ego had some doubts about the railroad's promise to fill in the Slough of Despond that had resisted efforts from time immemorial, Mr. Smooth-it-away pointed to a "convenient bridge" and explained:

> We obtained a sufficient foundation for it by throwing into the Slough some editions of books of morality; volumes of French philosophy and German rationalism; tracts, sermons, and essays of modern clergymen; extracts from Plato, Confucius, and various Hindu sages, together with a few ingenious commentaries upon texts of Scriptures—all of which, by some scientific process, have been converted into a mass like granite. The whole bog might be filled up with similar matter.

Thus sometime around the middle of the nineteenth century the work of engineers was beginning to be seen even by the layman to involve "some scientific process" as it transformed classical thinking into hard calculations. As engineering began to apply the scientific method to structural problems, it moved away from purely aesthetic considerations and separated itself from architecture. The roots of the two cultures' debate spread rhizoid-like in all directions throughout the Victorian era and periodically came to the surface here and there like dandelions in a spring lawn. Yet what is a weed and what a flower remains as difficult a problem of taxonomy as it was in Wordsworth's time.

As engineering came to mean the application of the scientific method to railroad bridges and other ambitious structures, its practitioners had to address the question of structural failure and structural success more explicitly. The failures of pyramids and cathedrals were by and large failures during construction, not failures during use. The failure of a railroad bridge was more likely to involve the lives not only of construction workers engaged in a high-risk activity but of innocent people who had entrusted their safety to the engineers. Sudden and catastrophic bridge collapses were introduced into the daily way of life, and they had to be reckoned with not through the classical trial and error method, but with a newer and more abstract method that employed pencil and paper in lieu of chisel and stone. What the engineers of the nineteenth century developed and passed down to those of the twentieth was the trial and error of mind over matter. They learned how to calculate to obviate the failure of structural materials, but they did not learn how to calculate to obviate the failure of the mind.

No one *wants* to learn by mistakes, but we cannot learn enough from successes to go beyond the state of the art. Contrary to their popular characterization as intellectual conservatives, engineers are really among the avant-garde. They are constantly seeking to employ new concepts to reduce the weight and thus the cost of their structures, and they are constantly striving to do more with less so the resulting structure represents an efficient use of materials. The engineer always believes he is trying something without error, but the truth of the matter is that each new structure can be a new trial. In the meantime the layman, whose spokesman is often the poet or the writer, can be threatened by both the failures *and* the successes. Such is the nature not only of science and engineering, but of all human endeavors.

6
DESIGN IS GETTING FROM HERE TO THERE

Designing a bridge or any other large structure is not unlike planning a trip or a vacation. The end may be clear and simple: to go from here to there. But the means may be limited only by our imaginations.

Let us imagine that we are living in Chicago and that we have promised to take our children to see New York City during two weeks of summer vacation. One of our first decisions is how to get to New York and back, and we generally can quickly narrow down the ways to three or four. We can drive our own car or take a bus or a train or a plane. The possibilities of going by hot-air balloon or bicycling or even taking a boat through the Great Lakes and man-made seaways will probably not occur to the average family, though they might to those who are so devoted to ballooning or biking or boating that the means of transportation may be considered more important than the destination itself.

Most families will end up choosing between driving and flying to New York, and they will base their decision on economic considerations (driving will be cheaper for larger families), convenience (driving will enable them to make their own schedule and also give them transportation about New York), aesthetics (driving will enable them to enjoy the scenery along the interstate), emotions (driving makes them less anxious than flying), or even habit (they always drive). Another family may use the same criteria to choose flying over driving because airfares are cheaper than tolls and fuel for their gas-guzzler, because having a car in New York will be a pain, because they love to look at the clouds when flying, because they know it is statistically safer to fly than to drive, and because they always fly. It is clear that there is no best way to get to New York and back, for the right choice for one family might clearly be the wrong one for another.

Even when the major decision has been made, there remain countless other minor decisions. What time should they leave? If they are driving, what route should they take? Should they drive straight through or should they stop overnight en route? Where should they stop? What motel should they stay in? Where should they eat dinner? If they are flying, how should they get to the airport? What flight should they take? How should they get from the New York airport (which one?) to their hotel (which one?)?

Clearly the choices can appear endless. Most choices will not make much of a difference as far as the principal object of spending about two weeks in New York is concerned, and thus they are not worth dwelling upon for an inordinate amount of time. Other decisions, such as whether to look for an inexpensive motel in New Jersey and drive into the city each day or whether to stay in a midtown hotel and be within walking distance of what one wants to do in New York, may have a major impact on whether the vacation is indeed a vacation.

All of the decisions about how to have a vacation in New York can be seen as attempts to maximize each family's enjoyment of its own vacation. Instead of stating its choices as preferences of one choice over another, each family might also reach its decisions by stating what it will not or does not want to do. Thus the large family might reject flying as too expensive, leaving them without a car in New York, depriving them of the opportunity to enjoy the scenic drive, introducing unnecessary anxiety into their vacation, and keeping them from their summer ritual of driving somewhere. Indeed, such a family might further insure the success of their vacation by reminding themselves *not* to go to the restaurant that disappointed them last time, *not* to wait in those long lines in Times Square for theater tickets, and *not* to try to drive crosstown during the rush hour. In short, the family can improve the success of its vacation by anticipating what can go wrong to ruin it.

Engineering design is not much different. Many objects of design are no more exotic than spending two weeks in New York. Even if you and your own family have not done it before, there are plenty of people willing to give you advice about what to do and what to avoid. There are books on the general subject of New York and others on such specialized aspects as the city's museums, restaurants, and shopping opportunities. Magazines and newspapers contain the experiences of the latest travelers to the Big Apple, telling you to be sure not to miss this or that attraction. And the availability and price of hotel rooms, theater tickets, and restaurants can be obtained over the telephone. In short, there is a wealth of experience and information out there for the asking. One can even find among his friends and neighbors experts on the negative aspects of visiting New York—crowds, the con men, the perverts, the muggers. Whether the small risk of encountering any of these detracts sufficiently from all the benefits of a visit is a subjective judgment that each family must make for itself.

The engineer designing a new highway bridge also has a wealth of experience available to him, as we all can imagine, having ourselves driven over and under tens of thousands of individual bridges on and off the interstate. We know there are still some covered wooden bridges out there, but today by and large we think in terms of concrete and steel. So many of the bridges resemble each other so closely that we soon pay no more attention to them than we do to the individual trees along the side of the road. But every now and then we come across a bridge that we can sense is special: a tall, curving arch spanning a deep ravine, a great suspension bridge across a wide bay, or a new cable-stayed bridge across a great river. Other bridges are special in ways that are not always obvious to the casual traveler, and sometimes these come to our attention only after their dramatic collapse. In some cases those colossal errors occurred because the bridge designer was doing what had not been done before, much as someone taking too exotic a holiday may not be able to arrange it through a conventional travel agent. To go through with exotic plans is clearly to be adventuresome, and one can ensure his safe and satisfied return by anticipating all that might go wrong. As the first trips of the astronauts to the moon demonstrated, travel to places without benefit of previous experience need not be doomed to failure.

The evolution of bridges can be traced back to primitive man felling a log across a brook, and the proud history includes the Roman aqueducts. But modern bridges are made neither of available logs nor of piles of stones. They are deliberate designs in concrete and steel arranged to suit the functional, aesthetic, and economic demands of our complex society. Because new demands are constantly being made—for a larger, more attractive, or less expensive bridge—it is not always possible, even if desirable, for the designer merely to copy what has been done successfully before. Copying may work for an ordinary highway bridge, but it clearly will not do when the highway is to cross a wider bay or a deeper ravine than ever spanned before. Then there are no examples to copy; there is no proven experience to follow. Thus the history of modern bridges is also the history of the development of a more scientific approach to designing large engineering structures than the pyramids and cathedrals or even the Roman aqueducts.

In 1779 the first iron bridge was erected in England at Coalbrookdale near the foundries that cast its iron arches with details that mimicked then-familiar (and successful) stone-and-timber construction. Ironbridge spanned a hundred feet across the Severn River and is still used by pedestrians today. At the time of its making, Ironbridge represented a bold experiment with a new material for bridge building, and it worked because the stone arch bridges on which its structural details were based had worked. But once iron had proven itself as a viable new construction material at Coalbrookdale, it was to be called upon to bridge ever wider gaps not only in space but also in knowledge. Because the new material could resist being pulled apart in ways that stone construction never could, new bridge designs sought to exploit iron in new ways. This combination of newly tried elements should have been a sure signal that the accumulated experience of centuries of stone would not be able to play a guiding role in the development of iron bridges, and it is not surprising that iron bridge makers would have to go through a period of trial and error to learn from their own and others' experiences.

The development and expansion of the railroads in the nineteenth century required bridges, and timber provided the material of many of the early railroad bridges. It was a familiar material and one that was usually available near the construction site. However, timber bridges required maintenance to be sure they did not rot away, and they were susceptible to the very fire that the iron horses carried in their bellies. It was inevitable that iron was to become a natural replacement for wood in bridges, but the conversion took a good part of the nineteenth century, in large part because iron bridges seemed not only new but also unpredictable. They collapsed in numbers that are still debated today.

In 1847 Queen Victoria appointed a commission to look into the use of iron in bridges, charging its members to "endeavor to ascertain such principles and form such rules as may enable the Engineer and Mechanic, in their respective spheres, to apply the Metal with confidence, and shall illustrate by theory and experiment the action which takes place under varying circumstances in Iron Railway Bridges which have been constructed." What was happening was that the bridges were experiencing what we know today to be fatigue failures, collapsing without warning under a passing train. The commission's report, published in 1849, promulgated a misguided idea about the fatigue of metals, theorizing that "crystallization" occurred under vibratory action, and this misconception was to persist well into the twentieth century. Nevertheless, the report did lead the English Board of Trade to formulate some requirements regarding stresses in bridge construction and thus had a beneficial effect on safety, even if it provided an inaccurate explanation of fatigue.

A considerable number of all-metal bridges were likewise built in America in the 1840s, but, in 1850, when a bridge in Pennsylvania broke under a train, all the metal structures on the New York and Erie Railroad were ordered replaced with wooden bridges. Not surprisingly, however, the development of the new technology persisted, with competing bridge designers and builders promising advantages in performance and price, and new iron bridges continued to be built. Despite the apparent risks, they came to be regarded as superior to wooden structures in the second half of the nineteenth century, in part because of the increasing availability of iron at decreasing cost, thus making it competitive with timber, which in some places was becoming more scarce and thus more costly.

No history of bridges is complete without at least an acknowledgement that many ambitious designs did fail. The famous ones, such as the Tay Bridge in 1879, the first Quebec Bridge failure in 1907, and the Tacoma Narrows collapse in 1940, are always mentioned, but the numerous failures of unnamed railroad bridges during the nineteenth century are generally grouped together as statistics. The casual mention in a recent book review that at one time iron truss bridges were failing at the rate of one in four elicited a series of letters to *Technology and Culture,* the international quarterly journal of the Society for the History of Technology, disputing the claim and offering various references and statistics of their own. The argument seemed especially curious because it pitted several humanists against one another in a debate over *numbers.* To an engineer, exactly how many railway bridges failed during the nineteenth century is not so important as the fact that failures did occur. The collapse of a *single* bridge made from a relatively new material or design should have been enough to make engineers of that era and their customers, the railroads, reflect on the new technology. The *repeated* collapses of iron railway bridges could only have cast suspicion on the adequacy of technological understanding and raised doubts among insiders and the public at large about the developing railroad industry. And doubts were raised.

One incident, the collapse of a bridge at Dixon, Illinois, prompted the American Society of Civil Engineers to create a committee to determine the most practicable means of averting such accidents. Though the committee was divided in its opinion, its report in 1875 made recommendations not only for railroad but also for highway bridge construction. Another specific incident, the collapse of the 157-foot-long truss bridge at Ashtabula, Ohio, in 1876, which killed almost a hundred people, prompted *Harper's Weekly* to ask, "Is there *no* method of making iron bridges of assured safety?"

As the nineteenth century drew to a close, the situation in Britain was apparently no different, as a macabre pencil drawing by John Tenniel, the illustrator of *Alice's Adventures in Wonderland,* suggests. His drawing of Death straddling the cracked girder of a railway bridge and captioned "On the Bridge!" appeared in an 1891 issue of *Punch.* The image was used to illustrate an anonymous essay with the same title, which, as its introduction stated, was meant to be a modernized version of Joseph Addison's "The Vision of Mirzah" (which had appeared in a 1711 issue of *The Spectator*). In Addison's allegory, the narrator Mirzah is led by a genius to a precipice above the valley of time that is spanned by a bridge, which represents human life, linking the beginning of the world and its end. The bridge Addison's Mirzah observes is in a ruinous condition, and many attempting to cross fall through. Mirzah passes some time marveling at the wonderful structure, but his heart fills with "a deep melancholy to see several [human figures] dropping unexpectedly in the midst of mirth and frivolity, and catching at everything that stood by them to save themselves."

Addison goes on to moralize, and so in his own way does the anonymous *Punch* parodist. In that modernized version, the protagonist, Matthew, contemplating "the Vanity of human Holiday-making," falls asleep over the pages of the multitudinous timetables of *Bradshaw's Monthly Railway Guide*. He dreams of being taken by a Genius to a pinnacle above "the Vale of Travel" traversed by "the great Railway System," which rises out of and recedes into the mists of "Monopoly and Muddle." The Genius directs the nineteenth-century Mirzah's attention to a metal bridge, of which Matthew reports:

> I found that the arch thereof looked shaky and insecure; moreover, that a Great and Irregular-shaped Cleft or Crack ran, after the fashion of a Lightning-flash in a Painted Seascape, athwart the structure thereof from Keystone to Coping. As I was regarding this unpleasing Portent, the Genius told me that this Bridge was at first of sound and scientific construction, but that the flight of Years, Wear and Tear, vehement Molecular Vibration, and, above all, Negligent Supervision, had resulted in its present Ruinous Condition. . . .
>
> Only an Attent, and, as it were, complacently Anticipative Visage, of an osseous and ogreish Aspect, gleamed lividly forth therefrom, as the Apparition appeared to Look and Listen through the Mist at one end of the Bridge for the welcome Sight of Disaster, the much desired Sound of Doom. A shrill and sibilant Metallic Shriek seemed to cleave the Shadows into which the Spectre gazed; a Violent Vibratory Pulsation, as of thudding iron flails threshing upon a

resonant steel floor, seemed to beat the Roadway, shake the
Bridge, and as it appeared to me to widen the levin-like
Crack which disfigured the Arch thereof.

When Matthew asks the significance of the Spectre, the Genius
replies gravely that it is "Insatiable Death waiting for Inevitable
Accident." When Matthew asks about Monopoly and Muddle,
the Genius is no longer found present. Matthew soon awakens and
changes his mind about a "Railway Excursion to Rural Parts" for
Holiday. Instead he goes to the pub and passes the day "in Safety
—and Solitary Smoking!" (an as yet unacknowledged Victorian
health hazard). The *Punch* story ends: "Next morning, however,
I read something in the papers which led me to believe that
Railwaydom Aroused meant exorcising and evicting that Sinister
Spectre, 'regardless of Cost'; and I shall look forward to my next
Holiday Outing with a mind Relieved and Reassured."

This suggests that even the railway companies themselves
recognized the unacceptable frequency of bridge failures, but what
the exact number was is a historical detail. Regardless of the
numbers, the *Punch* parody and the drawing by Tenniel—like
Nathaniel Hawthorne's earlier short story alluding to unsafe
bridges on the Celestial Railroad—demonstrate that the *perceived*
risk of failure, and not only of iron truss bridges, was certainly
high.

Our own contemporary technological failures perhaps bring
this point home more forcefully. It took only a single DC–10 crash
in Chicago in 1979 to ground the whole fleet, and the tragedy of
the collapsed walkways in the Kansas City Hyatt Regency Hotel
in 1981 was not diminished because it was a unique accident.
Dismissing the single structural failure as an anomaly is never a
wise course.

The failure of any engineering structure is cause for concern,
for a single incident can indicate a material flaw or design error
that renders myriad apparent structural successes irrelevant. In

engineering, numbers are means, not ends, and it ought rightly to have taken the failure of only a single bridge to bring into question the integrity of every other span. This elementary observation was accessible to monarch and commoner alike in the nineteenth century, as the record of that period shows. Today we speak of "technological literacy" and the need for non-engineers to be able to understand the ways and methods of technology. Neither Queen Victoria nor the nineteenth-century railroad traveler seemed to feel intellectually timid in the face of the technical issues of their day, and their example is a good one for twentieth-century citizens. While some of the details of engineering may be arcane, the principles of design and safety, of risk and benefit, are not, for to build a bridge is no less a human endeavor than to take a trip. The common expectation of engineer and layman is that the road will not lead to bridges that collapse.

Which innovation leads to a successful design and which to a failure is not completely predictable. Each opportunity to design something new, either bridge or airplane or skyscraper, presents the engineer with choices that may appear countless. The engineer may decide to copy as many seemingly good features as he can from existing designs that have successfully withstood the forces of man and nature, but he may also decide to improve upon those aspects of prior designs that appear to be wanting. Thus a bridge that has stood for decades but has developed innocuous cracks in certain spots may serve as the basis for an improved design of a bridge of approximately the same dimensions and traffic requirements. Or an existing design that has suffered no apparent distress after years or decades of service may lead the engineer to look for ways to make it lighter and thus less expensive to build, for the trouble-free prototype appears to be overdesigned.

The choices of design are ultimately like the choices of life. While the engineer can pursue on paper two or even many different designs that fulfill the requirements of a projected structure, in the last analysis only one design can be chosen to be built, just as, finally, only one route can be taken on a single trip from Chicago to New York no matter how many are considered in the planning. Deciding which paper design will be cast in concrete presents the designer or the selection committee with a problem not unlike that faced by Robert Frost:

> *Two roads diverged in a yellow wood,*
> *And sorry I could not travel both*
> *And be one traveler, long I stood*
> *And looked down one as far as I could*
> *To where it bent in the undergrowth;*

The designer tries to look ahead to determine where his different designs will end up down the road, but the way through the future always seems to fork again and again and to become fuzzy in the undergrowth. An ultra-conservative engineer may take the path that others have taken and opt for the familiar bridge, even though it is more expensive than a newer design or despite its unattractiveness for the site. A bold and imaginative engineer can infuse a bridge with meaning, as Frost did a common fork in the road. Robert Maillart, the Swiss engineer who has been called a structural artist, developed innovative concrete designs of both economy and beauty that revolutionized bridge construction early in this century, and he might be said to have followed the poet:

> *Two roads diverged in a wood, and I—*
> *I took the one less traveled by,*
> *And that has made all the difference.*

7

DESIGN AS REVISION

There is a familiar image of the writer staring at a blank sheet of paper in his typewriter beside a wastebasket overflowing with crumpled false starts at his story. This image is true figuratively if not literally, and it represents the frustrations of the creative process in engineering as well as in art. The archetypal writer may be thought to be trying to put together a new arrangement of words to achieve a certain end—trying to put a pineapple together, as Wallace Stevens said. The writer wants the words to take the reader from here to there in a way that is both original and familiar so that the reader may be able to picture in his own mind the scenes and the action of the story or the examples and arguments of the essay. The crumpled pages in the wastebasket and on the floor represent attempts that either did not work or worked in a way unsatisfying to the writer's artistic or commercial sense. Sometimes the discarded attempts represent single sentences, sometimes whole chapters or even whole books.

Why the writer discards this and keeps that can often be attributed to his explicit or implicit judgment of what works and what does not. Judging what works is always trickier than what does not, and very often the writer fools himself into thinking this or that is brilliant because he does not subject it to objective criticism. Thus manuscripts full of flaws can be thought by their authors to be masterpieces. The obviously flawed manuscripts are usually caught by the editor and sent back to the author, often with reasons why they do not succeed. Manuscripts that come to be published are judged by critics and the general reader. Sometimes critic and reader agree in their judgment of a book; sometimes they do not. Positive judgments tend to be effusive and full of references to and comparisons with other successful books; negative judgments tend to be full of examples demonstrating why the book does not work. Critics often point out inconsistencies or infelicities of plot, unconvincing or undeveloped characters, and in general give counter-examples to the thesis of author, editor, and publisher that this book works. In short, the critic points out how the book *fails.*

The point was made quite explicitly in a recent review of *The Man Who Could See Through Time,* a play by Terri Wagener in which a physics professor and a young sculptor debate science and art. Reviewing the play in *The New York Times,* Frank Rich wrote:

> The best two-character plays look so effortless that we tend to forget how much craft goes into them. . . . To see a two-character play that fails, however, is to appreciate just how difficult the form really is.

And so it is with engineering structures. The great suspension bridges look so simple in line and principle, yet the history of failures of the genre has demonstrated that their design takes a touch of genius. And geniuses like Washington Roebling and Othmar Ammann can arguably be said to have learned more what not to do from the great failures of their forgotten predecessors than today's designers can be expected to learn about how to design the next suspended masterpiece from either the Brooklyn or the Verrazano Narrows Bridge.

Some writers, even if they do not try to publish them, do not crumple up their false starts or their failed drafts. They save every scrap of paper as if they recognize that they will never reach perfection and will eventually have to choose the least imperfect from among all of their tries. These documents of the creative process are invaluable when they represent the successive drafts of a successful book or any work of a successful writer. What other authors tend to learn from the manuscripts and drafts of the masters that cannot be learned from the final published version of a work is that creating a book can be seen as a succession of choices and real or imagined improvements. An opening sentence or even a word may evolve to its final form only after going through dozens of rejected alternatives. Sometimes the final version is closer to the first than any of the intervening versions, and sometimes a word will be crossed out only to have a ladder of synonyms, near synonyms, remote synonyms, and even antonyms leading like Jack's beanstalk through the clouds of imagined riches, ultimately to have the author fall back on the very word with which he began. These creative iterations suggest that the author's choice of even a single word is more easily understood in terms of rejection than acceptance, in terms of failure rather than success, in terms of *no* rather than *yes*. The fair manuscript gives little if any hint as to why exactly the author put down what he did in his first draft. But the word changed, the sentence deleted, and other alterations that may be traced through successive drafts show clearly that the author did not believe what he had originally written had been right. It failed in some way to contribute to the end that the author was working toward. This is not to say that something unchanged from first to last was deemed perfect by the author; it simply indicates that, rightly or wrongly, he detected no unacceptable fault with it or could see no alternative. The unchanged part of his book's structure might teach the student nothing about its composition, however.

Some of the acknowledged masters of the written word were seemingly never completely satisfied with their work. James Joyce was apparently notorious for making voluminous changes even as his major works *Ulysses* and *Finnegans Wake* were being set in type by the printer. And what was set in type was revised by Joyce in the proofs. In 1984, after volumes of criticism were published based on the original 1922 version of *Ulysses,* a new edition appeared, reportedly correcting over five thousand errors that crept into the first edition. The book was claimed by one critic to be so changed by the restoration of a few dropped lines that a whole new interpretation of the novel was in order.

Other recognized masters often express the thought that they *abandon* a work rather than complete it. What they mean is that they come to realize that for all their drafts and revisions, a manuscript will never be perfect, and they must simply decide when they have caught all its major flaws and when it is as close to perfect as they can make it without working beyond reasonable limits. Even the twenty-odd years Joyce spent on *Finnegans Wake* was apparently not enough for him to believe he gave a perfect manuscript to the printer, and all authors acknowledge implicitly that revisions to manuscripts reach a point of diminishing returns.

The work of the engineer is not unlike that of the writer. How the original design for a new bridge comes to be may involve as great a leap of the imagination as the first draft of a novel. The designer may already have rejected many alternatives, perhaps because he could see immediately upon their conception that they would not work for this or that reason. Thus he could see immediately that his work would fail. What the engineer eventually puts down on paper may even have some obvious flaws, but none that he believes could not be worked out in time. But sometimes even in the act of sketching a design on paper the engineer will see that the approach will not work, and he crumples up the failed bridge much as the writer will crumple up his abortive character sketch.

Some designs survive longer than others on paper. Eventually one evolves as *the* design, and it will be checked part by part for soundness, much as the writer checks his manuscript word by word. When a part is discovered that fails to perform the function it is supposed to, it is replaced with another member from the mind's catalog, much as the writer searches the thesaurus in his own mind to locate a word that will not fail as he imagines the former choice has. Eventually the engineer, like the writer, will reach a version of his design that he believes to be as free of flaws as he thinks he can make it, and the design is submitted to other engineers who serve much as editors in assessing the success or failure of the design.

As few, if any, things in life are perfect, neither is the analogy between books and engineering structures. A book is much more likely to be an individual effort than is a building, bridge, or other engineering structure, though the preparation of a dictionary or an encyclopedia might be said to resemble the design of a modern nuclear power plant in that no individual knows everything about every detail of the project. Furthermore, the failure of a book may be arguable whereas the failure of a building collapsed into a heap of rubble is not. Yet the process of successive revision is as common to both writing and engineering as it is to music composition and science, and it is a fair representation of the creative process in writing and in engineering to see the evolution of a book or a design as involving the successive elimination of faults and error. It is this aspect of the analogy that is most helpful in understanding how the celebrated writers and engineers alike learn more from the errors of their predecessors and contemporaries than they do from all the successes in the world.

While engineers who play it safe and copy designs that have stood the test of time may be well paid (though perhaps not nearly as well paid as the authors of mass market paperbacks who use more formulas in their genre fiction than are available to any engineer), there is no more professional distinction in being such an engineer than there is literary recognition in being such an author. When we speak of creative engineers we are talking about as select a few as there are great writers. And just as the great writers are those who have given us unique and daring experiments that have worked, so it is that the great engineers are those who have given us their daring and unique structural experiments that have stood the test of time.

It is not capricious to compare engineers with artists on the one hand and with scientists on the other. Engineering does share traits with both art and science, for engineering is a human endeavor that is both creative and analytical. Being creative pursuits, the innovative works of engineering test the vocabulary of the critics, whoever they may be, and it is not always clear-cut whether a daring new structure will stand or fall, even in the make-believe world of hypothesis testing. The problem with the new structure lies in the very humanness of its origins and of the world in which it will function. Whether an engineering structure succeeds as a work of art may not be a question of life and death, but whether it will stand or collapse beneath the weight of those attending the dedication ceremonies is indeed a question to be reckoned with.

The very newness of an engineering creation makes the question of its soundness problematical. What appears to work so well on paper may do so only because the designer has not imagined that the structure will be subjected to unanticipated traumas or because he has overlooked a detail that is indeed the structure's weakest link. Certainly no designer who remembers the ill-fated Tacoma Narrows Bridge will design another bridge like it, but a new bridge that is unique aesthetically or analytically may hold surprises even for the designer. To be safe the engineer should try to imagine the structure in every conceivable situation and check each case to be sure that not even the slightest part is liable to break. But to imagine and check *every* conceivable situation might take forever, and the engineer must make judgments as to which situations are the most severe and which are insignificant. The former are analyzed, while the latter are ignored.

But as the literary critic can discover meanings and symbols an author denies having been aware of in a piece of creative writing, so the analytical critic of a new engineering design can find interactions among the parts of a structure that surprise the designer. And just as a literary reevaluation can come years after a book has achieved critical acclaim, so an engineering structure can stand, though precariously, for years before it is reanalyzed, perhaps because it or a structure not unlike it has recently failed. In lucky cases the faulty structure will be caught before it collapses; in many cases it is catastrophe that prompts the postmortem exposé.

The engineering task of designing a bridge shares qualities with the tasks of both poetry and science. Like poetry, the exact bridge one designer conceives to span a given space during a given technological era may never exist in the mind of any other engineer at any other time. Yet, like discoveries in science, if the theoretical and motivational foundations for a bridge are laid, then a bridge will be built, and it will be *the* bridge for that place and time no matter who designs it. No poetic license is allowed in the design of the details of the bridge, for an erratic line on the blueprint or an eccentric number in a calculation can be the downfall of the structure, no matter how much like a sound bridge it looks on paper. And today, if a computer is used, even so small an error as an inadvertent slip of a punctuation mark, decimal point, or sign in an equation can lead to a bridge that fails even if the computer model works.

The bridges of Robert Maillart have been praised as works of art that are one with the Alps. In flights of innovation the Swiss engineer may be thought to have spanned the intricacy of contour lines on maps over which his mind floated like an eagle, light to the eye but strong to the touch. Like the pensive man in Wallace Stevens' "Connoisseur of Chaos":

> . . . He sees that eagle float
> For which the intricate Alps are a single nest.

Maillart's use of concrete reinforced with steel put poetry in place of prosaic bulk, whose ease of understanding may have made a more conventional bridge clearly safe but did not distinguish it. Maillart's bridges were not conceived fully justified line by line upon the page, for they work as entities as spare as poems. And as with poems, in the end, they work simply because they work. Upon their completion on paper, Maillart's plans could be analyzed after a fashion, and they could be somewhat revised here and there to redistribute stresses or smooth the lines. Their proof was in the putting the plans into place, however. And when the concrete had set about the steel, then the falsework supporting the bridge a-building (as the platen supports the poem a-typing) was removed, and each bridge withstood its first test as the poem its first reading. As each new bridge endured the seasons of use and re-use, as the poem the readings and re-readings, each bridge became a success. But Maillart also saw in many of his master-pieces extra weight or an unnecessary line that he removed in the next. His practice of self-criticism and revision was not unlike the writer's.

The engineer no less than the poet sees the faults in his crea-tions, and he learns more from his mistakes and those of others than he does from all the masterpieces created by himself and his peers. While Maillart mastered the steel-in-concrete bridge, he did not originate the form. In the closing years of the nineteenth century François Hennebique, whose French construction firm restored Gothic cathedrals among its other activities, carried out research on the use of steel embedded in concrete to resist the cracking that invariably occurs when concrete is tensed rather than compressed, as it is in all-masonry structures. Hennebique's firm gained experience in reinforced concrete structures by com-pleting thousands of projects, but these were not all without flaws. The bridge crossing the Vienne River at Chatellerault, France, in particular, developed a number of cracks. This was the longest Hennebique arch bridge, and its pattern of cracking showed to all who would observe it that there was room for improvement in the design. As David Billington has argued in his award-winning monograph on Robert Maillart's bridges, Maillart, in particular, learned from such experiences, as he learned from the cracking of his own designs, such as his early bridge at Zuoz, Switzerland.

Small cracks in reinforced concrete do not necessarily pose any danger of structural collapse, for the steel will resist any further opening of the cracks. But cracks do signify a failure on the designer's part to understand his design completely when it was only on paper, since the cracks incontrovertibly disprove the implicit hypothesis that stresses high enough to cause cracks to develop would not exist anywhere in the structure. Discovering cracks in a completed structure thus enables the designer to learn the weaknesses of his knowledge and thereby to improve upon future designs, if only to beef them up where the stresses are less precisely predictable. In this way an engineer's designs can evolve from early works that show promise to the mature works that are brilliant and develop no cracks, much as a poet's juvenilia evolve into masterpieces that appear to be seamless.

Engineering, like poetry, is an attempt to approach perfection. And engineers, like poets, are seldom completely satisfied with their creations. They notice, even if no one else does, the word that is not quite *le mot juste* or the hairline crack that blemishes the structure. However, while poets can go back to a particular poem hundreds of times between its first publication and its final version in their collected works, engineers can seldom make major revisions in a completed structure. But an engineer can certainly learn from his mistakes.

Anton Tedesko, designer of such significant concrete shell structures as the sports arena in Hershey, Pennsylvania, and the airport terminal in St. Louis, relates the story of inspecting fine hairline cracks that developed in a hyperbolic paraboloidal shell in Denver that he designed for I. M. Pei. The concrete shell does not possess the stiffening ribs that have been criticized as architectural blemishes on the St. Louis airport shell, but Tedesko believes the hypar shell's cracks, which he suspects most people are not even aware of, could have been avoided and he has inspected them for over two decades, whenever he stops in Denver. It is not that an engineer admires his own or gloats over others' mistakes, it is that he recognizes, unfortunate though they may be, that defects are unplanned experiments that can teach one how to make the next design better. As Tedesko has written in a retrospective article on concrete shell structures that interweaves stories of collapses with stories of triumph: "I have mentioned examples of unfortunate experiences because it is easier to draw lessons from examples of poor performance than from good performance." Art and literary criticism serve the same purpose, and it is not surprising that artists and writers, like engineers, are often their own severest critics.

Engineering students understand early on that there is a great deal to be learned from a mistake. In a recent engineering class at Duke, students participated in the design of an experiment to produce metals through a foaming process aboard the space shuttle to test the feasibility of manufacturing new lightweight structural materials in a weightless environment. One student in the course admitted that "real engineering is a lot harder than we thought," but he also recognized that the class was "learning a lot from failing and screwing up." These students were coming to the realization that T. H. Huxley expressed in his book *On Medical Education:* "There is the greatest practical benefit in making a few mistakes early in life."

8
ACCIDENTS WAITING TO HAPPEN

On Friday evening, July 17, 1981, the lobby of the newest hotel in Kansas City was crowded with dancers and others just enjoying the big band sound swinging them into another weekend. Many watched from architecturally graceful walkways suspended across the grand expanse of the columnless atrium, and many tapped their toes and swayed to the strains of "Satin Doll." The gentle undulations of the walkways themselves only added to the general celebration of the evening, which was to end with the catastrophic collapse of two crowded walkways onto the even more-crowded floor below. Thus the Kansas City Hyatt Regency Hotel became a synonym for the greatest structural tragedy in the history of the United States, and grieving, accusations, investigations, and lawsuits would continue for years after the accident.

There were 114 people killed and almost 200 others injured, and it was estimated that approximately half of Kansas City's population was directly or indirectly touched by the tragedy. Survivors everywhere wondered how such a terrible thing could have happened, and the national news media covered the incident intensely for several days. But soon the story was driven off the front page and received less and less notice, except in the Kansas City newspapers, which of course had a more immediate interest in the story. The newspapers engaged consulting engineers to review the evidence and advise their readers of the causes of the accident, and within four days of the structural failure, the front page of *The Kansas City Star* carried in lieu of headlines technical drawings of design details that pinpointed the cause. Investigative reporting that would win the Pulitzer Prize identified the skywalks' weak link and explained how a suspension rod tore through a box beam to initiate the progressive collapse of the weakened structure. A deviation in design in the way the rods connected the lower skywalk to the upper and the upper to the ceiling of the atrium was clearly described and zeroed in on as the ultimate cause of the accident.

Architectural drawings for the original design of the suspension system for the hotel's skywalks showed at each point of support a single long rod attached to the ceiling and passing through the cross beams on which the walkways rested. The two walkways were hung one below the other by washers and bolts under their respective floor beams. This original design could work in principle if the long rods were strong enough, the washers and bolts large enough, and the floor beams heavy enough to support not only the walkways themselves but also the crowds of people that might be expected to assemble and perhaps even to run, jump, or dance along them. Precisely what "strong enough," "large enough," and "heavy enough" translate into in terms of numbers representing engineering stresses and strains is normally governed by building codes, but such codes apparently did not explicitly treat a unique structural concept like the Hyatt Regency skywalks.

But even if they do not specifically address the design of skywalks, building codes do require a certain level of strength for steel rods in tension and do specify such assumptions as how much weight people can be expected to place upon hotel walkways, whether or not suspended from a ceiling or another walkway. By these standards, the original support system planned for the Hyatt Regency walkways was found by investigators at the National Bureau of Standards to be underdesigned. The supports would have been only sixty percent as strong as they should have been according to the Kansas City Building Code. However, since the writers of building codes want designs to be on the safe side and to provide plenty of margin for error not only in assumptions and calculations but also in steelmaking and construction, the codes require much more than minimum strength. Thus the walkways as originally designed, though not as strong as they should have been, would probably not have fallen, and their nonconformity with the code might never have been discovered.

Unfortunately, someone looking at the original details of the connection must have said he had a better idea or an easier way to hang one skywalk beneath the other and both from the sixty-foot-high ceiling of the atrium, for the connection on the original architectural drawing was difficult, if not impossible, to install. The original concept of a single rod extending from each ceiling connection, passing through holes in a beam supporting the fourth-floor walkway about fifteen feet below and continuing for another thirty feet before passing through the beam of the second-floor walkway, would indeed have been an unwieldy construction task, and someone's suggestion that two shorter rods be used in place of each long one must have had an immediate appeal to anyone involved with the erection of the walkways. (We all remember or can imagine how easily bent the longest pieces of our Erector sets were and how they never looked quite the same after we tried to straighten them out again.)

In the modified connection system, one support rod extended from the ceiling through holes in the top walkway's beam, which rested on a washer and nut threaded onto the end of the rod. Through another set of holes a few inches toward the inside of the upper walkway, a second rod was hung down, prevented from being pulled through by another washer and nut, and a lower walkway beam was supported from this rod in the same way the upper one was from the upper rod. This was certainly an easier way of assembling the skywalks, for it is a much more familiar task to thread rods at their ends and fasten nuts to them than to thread a forty-five-foot-long rod along its entire length or to cut threads only where the rods were to pass through the upper walkway's beam, as suggested by the original drawings. But no matter how much more convenient to assemble, the new rod configuration effectively doubled the push of the washer on the box beam supporting the upper walkway's floor, and this made the already underdesigned skywalks barely able to support their own weight.

We can appreciate the mechanics of the problem by considering a simple analogy, replacing the original rod with a rope and the two walkways with two people hanging onto the rope. Each person's weight is transferred to the rope through the hands gripping it. The support of the walkways in the original design is analogous to the two people hanging separately, one below the other, onto the same rope. If the rope were strong enough and the individuals' grips were tight enough, each person could hold onto the rope without falling. If, however, the lower rope-hanger grabs not the same rope but another tied to the legs of the person above, the upper person's grip must support two bodies, or roughly twice his own weight. It is not the strength of the rope that becomes critical, but the strength of the top person's grip. If his grip were easily able to support his weight but is just barely strong enough to support the doubled weight, his hands would tend to give out under the pull. Eventually, even the slightest wiggling of the lower person could loosen the upper's grip and send them both crashing to the ground.

In the wake of the hotel tragedy the National Bureau of Standards not only determined that the original design was not as strong as it should have been, but also found by testing box beam and rod connections—the details that effectively gripped the rod —that the skywalks as constructed were just barely able to support their own weight, thus confirming that the skywalks were an accident just waiting to happen.

There were apparently several incidents during construction of the Hyatt Regency Hotel when the weakness of the walkways could have been detected, but like the warnings of some eventual suicides, these signals of potential disaster were not heeded. The lobby roof collapsed when the hotel was under construction, and checks of many structural details including those of the skywalks were ordered, but apparently the rod and box beam connections were either not checked or not found wanting. After the walkways were up there were reports that construction workers found the elevated shortcuts over the atrium unsteady under heavy wheelbarrows, but the construction traffic was simply rerouted and the designs were apparently still not checked or found wanting.

Though it is easy to analyze in retrospect, the weak link in the Hyatt Regency elevated walkways somehow escaped notice amid the myriad other details of the unique construction situation. However, after the accident there were Monday-morning quarterbacks galore, and they were nowhere so much in evidence as in the letters-to-the-editor section of *Engineering News-Record.* The original details of the architectural drawings did not show exactly how the single long rod was to be manufactured, and letters in the wake of the accident made clear that at least some readers of that widely circulated weekly of the construction industry thought the original architectural drawing called for a part as impossible as those that would be needed for one of M. C. Escher's buildings. Indeed, what was published as the original drawing showed a nut at least fifteen feet up the single rod, and many readers apparently scratched their heads as to how that nut would get there. They pointed out that the whole length of rod between the nut and the end would have to be threaded, which the drawings did not show, or the threaded section of the rod would have to be thicker than the rest. One reader summed up the consensus: "A detail that begs a change cannot be completely without blame when the change is made."

Within a month of the letters pondering the "Hyatt Puzzle" of putting a nut fifteen feet up a rod, a second volley of letters appeared in *Engineering News-Record,* these being headlined, "A Hyatt Puzzle Solution." These readers did not take the architectural drawings literally and pointed out how a pair of not inconveniently long rods could be connected end to end by a single "sleeve nut"—one that effectively serves as a coupling device—to eliminate the need for offsetting the two rods and creating the dangerous condition that resulted. Another reader suggested a possible way in which the same steel channel sections that constituted the box beam could have been welded in a different configuration to provide a stronger bearing surface, and one can only wonder how many other unpublished suggestions the mailbag of *Engineering News-Record* held. When I discussed the Hyatt Regency failure in an article in *Technology Review,* the editor of that magazine also received an unusually large amount of mail on the walkway detail. Sleeve nuts, split dies, and a host of other solutions to the puzzle were proposed by readers, who apparently did not read *Engineering News-Record* and who seemed incredulous that such a detail caused so much trouble.

But explaining what went wrong with the Hyatt Regency walkways and pointing out changes that would have worked is a lot easier than catching a mistake in a design yet to be realized. After the fact there is a well-defined "puzzle" to solve to show how clever one is. Before the fact one must not only define the design "puzzle" but also verify one's "solution" by checking all possible ways in which it can fail. One wonders how many authors of the letters-to-the-editor would have caught the error in the walkway design had its ill-fated detail not been highlighted among the dozens of others, not only for the walkway but for all the other parts of the hotel. Before the rod and box beam assembly had failed it might not have been so conspicuous to the untrained eye.

The letters to the editor and many editorials and articles in the pages of professional engineering publications inspired by the Hyatt Regency accident suggested that it would not have happened were experienced designers and detailers involved. They would most likely have used a tried-and-true detail that they knew would work, or they would likely have looked very carefully at such an unusual one as specified for the walkway support. This kind of claim can never be verified, of course, but it is no doubt more likely for someone with a considerable amount of experience to be able to spot things that will not work. However, just as no one who knows of the Tacoma Narrows Bridge is likely to ignore the effect of wind on a suspension bridge, so no one who remembers the Hyatt Regency skywalks is likely to let another rod-beam connection escape close scrutiny. Thus the tragedy no doubt made a lot of inexperienced detailers suddenly much more experienced. And it is precisely to keep these lessons in the minds of young engineers that failures should be a permanent part of the engineering literature.

After twenty months of investigation, the U. S. attorney and the Jackson County, Missouri, prosecutor announced jointly that they had found no evidence that either a federal or a state crime was committed in connection with the Hyatt Regency skywalks collapse. However, within two months of that announcement, the attorney general of Missouri filed a complaint against the engineers responsible for the design of the hotel. They were charged with "gross negligence" in the design and analysis of the suspended walkways. Specifically, they were charged with failure to "perform calculations to determine the load capacity of the bridge rods and connections," failure to check if the rods specified would be adequate after the design changes were made, and failure to perform an analysis of the walkway suspension system as requested by the owner after a section of the hotel roof collapsed during construction. The law is no more an exact science than engineering, and the strength or weakness of the attorney general's case awaits the test of a trial as this is being written.

Once one of the hanger rods pulled through the box beam connection under the Hyatt Regency walkway, that rod clearly could no longer carry its share of the weight of the skywalks and the people on them. However, had the structure not been so marginally designed, the other rods might have redistributed the unsupported weight among them, and the walkway might only have sagged a bit at the broken connection. This would have alerted the hotel management to the problem and, had this warning been taken more seriously than the signs of the walkway's flimsiness during construction, a tragedy might never have occurred. Thus designers often try to build into their structures what are known as *alternate load paths* to accommodate the rerouted traffic of stress and strain when any one load path becomes unavailable for whatever reason.

When alternate load paths cannot take the extra traffic or do not even exist, catastrophic failures can occur. The Hyatt Regency skywalks were a kind of bridge across the hotel's atrium, and bridges are often daring structures because of the several functions they serve. In the case of the skywalks, there was an architectural effect of openness to be maintained in the great atrium lobby, which had come to be a trademark of Hyatt hotels. When the skywalks collapsed there was immediate concern about the safety of other daring lobby designs, but none was quite the same as the Kansas City concept. Chicago city inspectors ordered the restricted use of suspended areas of that city's Hyatt Regency until reinforcements could be made, but generally the effects of the Kansas City failure will be on new construction that might bear some resemblance to the ill-fated skywalks.

Within days of the skywalk tragedy a third, remaining elevated walkway that had been suspended alongside the two that collapsed was dismantled and removed from the lobby in the middle of the night despite protests from the mayor of Kansas City. The owners of the hotel argued that the remaining walkway presented a hazard to workers and others in the building, but attorneys for some of the victims objected, claiming that evidence would be destroyed, and engineers interested in studying the cause of the accident lamented the removal of the only extant structure anywhere near equivalent to that which had collapsed. Had the third walkway not been removed so precipitately, its behavior under the feet of dancers might or might not have confirmed the theory that the walkways responded to the tempo of the dance the way a wine glass can to a soprano's high note.

Today the Hyatt Regency lobby in Kansas City is spanned by a single walkway resting on stout columns sitting on the solid floor. While the open expanse of the lobby is perhaps diminished architecturally, the new design is no doubt intended to provide a sense of security and bear little resemblance to the fragile sky-walks. But even though the reconstructed walkway in the hotel lobby may be as safe as one can imagine, the apparent absence of a skyhook-like suspension system in other bridges has not meant that they were not in fact lacking alternate load paths to compensate for potential failures of the primary structural supports. Many bridges over busy waters have had sections knocked down when a heavy barge or ship collided with one of their piers, and some bridges that appear to be resting on columns really are not so simply constructed.

Approximately two years after the collapse of the Hyatt Regency walkways, a hundred-foot-long and three-lane-wide section of the Connecticut Turnpike over the Mianus River suddenly fell out of the elevated highway. Four vehicles plunged seventy feet to the river bank, with three persons killed and another three injured in the accident. There might easily have been more victims had the structure not failed in the early morning hours when traffic was light, for this section of Interstate 95 near Greenwich normally bears a hundred thousand vehicles a day. While the re-routed traffic caused some inconveniences and disturbed the peace of small towns in the vicinity, alternate routes could at least be established for the highway traffic between Connecticut and New York. But the bridge failed because there was no alternate load path with the capacity to support the rerouted weight of the structure when part of it became loose.

The Mianus Bridge was what is known as a "pin-connected hung span," complicated by the fact that it crossed the river at an angle and was thus skewed on its supports. Two of the four corners of the part that fell were hung from a part of the bridge that was cantilevered from the piers on which it rested. The means of hanging was much like the link of a bicycle chain, with one end pinned to a cantilevered girder and the other end to one of the girders of the hung span. These links were over six feet long and the connection was completed with seven-inch diameter steel pins that passed through the links and girders. The pins were designed to be held in place by a bolt that passed through a hole along the length of the pin and held large, thick washers close to the sides of the links.

In hearings conducted by the National Transportation Safety Board in the wake of the accident there seemed little doubt that the Mianus Bridge collapsed after one of the hanger links failed. Exactly how and why it failed was a matter of dispute, however. The skewed nature of the design was blamed by one consultant as introducing forces that caused one of the pins to work itself loose during the twenty-five years that the bridge was standing. Testimony from the firm that designed the bridge focused on improper maintenance as a contributing factor, pointing out that the bridge's drainage system was paved over, causing extensive corrosion to the critical links due to rainwater contaminated by road salt. State transportation officials acknowledged improper drainage from the structure, but they claimed miscalculations involving the hanger details were made by the designers.

The failure of the Mianus Bridge recalled that of the Point Pleasant Bridge across the Ohio River in 1967. This 1,750-foot-long bridge had carried traffic between Gallipolis, Ohio, and Point Pleasant, West Virginia, for thirty-eight years when it suddenly collapsed under approximately seventy-five vehicles during rush hour and caused forty-six deaths. The roadway of Silver Bridge (as the Point Pleasant Bridge was nicknamed due to the fact it was the first bridge to be painted with aluminum paint) was suspended not from the round wire cables familiar in most modern suspension bridges but from two giant chains made up of fifty-foot-long links, or "eyebars." The bridge was the first to use eyebars of a special high-strength steel, and these were connected bicycle-chain style with bolts and washers. The assembly not only made inspection difficult but also encouraged corrosion to occur throughout the years. Since this deterioration went undetected, small pits emanating from the holes in the steel links grew through the process of fatigue-crack growth accelerated by corrosion. When a crack had weakened one link to the extent that its strength was exceeded by the load imposed by the heavy traffic added to the weight of the bridge itself, the link broke. The load previously carried by the broken link had to be borne by other links, and the chain was twisted by the unbalanced geometry. This apparently caused other links to slip off their pins, and the load path to the bridge towers was broken. The bridge progressed rapidly to total collapse through want of something to hold on to.

The common feature of these failures—the Hyatt Regency skywalks, the Mianus River bridge, and Silver Bridge—lies in the details. The skywalk connections were not strong enough in the first place and the bridges literally contained weak links that became weaker with time. Can we generalize that all structures that cannot tolerate the loss of one of their connections should be dismantled immediately lest we be surprised with the failure of another detail that can lead to progressive collapse? That would mean that all DC–10 airplanes would have to be grounded as they were in 1979 in the wake of the crash of one upon takeoff from Chicago's O'Hare Field. The engines of the DC–10 are connected to the wing through a pin assembly, and the one that failed on the Chicago jumbo jet was supposed to function during takeoff in a manner similar to the pin-and-link assembly in a bridge under rush-hour traffic. What happened to the ill-fated DC–10 was that the engine was removed for servicing and replaced in a way unanticipated by the designer. Thus the flange into which the link (or "pylon" in aircraft terminology) fitted became torn around its pin-hole much as the ring-hole in a piece of loose-leaf paper becomes torn through being pulled too roughly or otherwise mistreated in the binder. If we do not want our important loose-leaf pages falling out of our binders, we can repair them with hole reinforcers as soon as we notice tears beginning to appear, or we can change our habits and turn the pages less forcefully. Depending on how important to us it is that not a single leaf falls from the binder, once we recognize the potential for leaves coming loose we can be as conscientious as we want to be about preventing it. If we use two-hole binders, we can replace them with three-hole designs to provide alternate load paths should one hole tear. Or if, as is likely, we already have three-hole binders, we can simply go through the binder every now and then and inspect the condition of each hole, reinforcing those that show an unacceptable amount of tearing. But it is unlikely that we would think much about reinforcing loose-leaf holes unless a page actually tore out or we noticed that one was about to.

Since the engine attachment assembly on the DC–10 was not suspected to be weakening before the fatal crash, no one paid any special attention to it. However, as soon as the cracked flange was identified as the cause of the Chicago crash, the flanges of all DC–10s, which had been grounded pending the discovery of the cause of the failure, were inspected. Cracks were found in several pylon assemblies, and eventually it became clear that the cracks were introduced by the excessive and unanticipated abuse of the flange during maintenance. With this weak link in the airplane's design identified, the DC–10s with no pylon assembly cracks were allowed to resume flying, those with cracks were repaired, and all pylon assemblies were henceforth maintained and inspected with special care. Furthermore, the maintenance procedure that led to failure was discontinued, and such techniques generally came under closer scrutiny. The DC–10 has continued to fly safely, but the lesson of the accident will not easily be forgotten.

Thus the lessons of failures generally pinpoint weak links. Existing structures with those same weak links can safely be allowed to stand or fly because knowledge is power over the built as well as the human population. The weak link can be avoided or strengthened in future designs, and the science of that genre of weak-linked structures can generally be said to have benefited in a way that years or even decades of skywalks hanging, bridges standing, or DC–10s flying did not. For this reason it is important that engineers study failures at least as much, if not more than successes, and it is important that the causes of structural failures be as openly discussed as can be. Should a young engineer look for models in weak-linked structures while they are still functioning, he could indeed design weak links into his own structures. However, if the cause of a failure is understood, then any other similar structures should come under close scrutiny and the incontrovertible lesson of a single failed structure is what *not* to do in future designs. That is a very positive lesson, and thus the failure of an engineering structure, tragic as it may be, need never be for naught.

9
SAFETY IN NUMBERS

While engineers can learn from structural mistakes what not to do, they do not necessarily learn from successes how to do anything but repeat the success without change. And even that can be fraught with danger, for the combination of good luck that might find one bridge built of flawless steel, well-maintained, and never overloaded could be absent in another bridge of identical design but made of inferior steel, poorly maintained or even neglected, and constantly overloaded. Thus each new engineering project, no matter how similar it might be to a past one, can be a potential failure. No one can live under conditions of such capriciousness, and the anxiety level of engineers would be high indeed if there were not rational means of dealing with all the uncertainties of design and construction. One of the most comforting of means, employed in virtually all engineering designs, has been the *factor of safety*.

The factor of safety is a number that has often been referred to as a "factor of ignorance," for its function is to provide a margin of error that allows for a considerable number of corollaries to Murphy's Law to compound without threatening the success of an engineering endeavor. Factors of safety are intended to allow for the bridge built of the weakest imaginable batch of steel to stand up under the heaviest imaginable truck going over the largest imaginable pothole and bouncing across the roadway in a storm.

While of course there will have to be a judgment made as to which numbers represent these superlatives, the objective of the designer is to make his structure tough rather than fragile. Since excessive strength can be unattractive, uneconomical, and unnecessary, engineers must make decisions about how strong is strong enough by considering architectural, financial, and political factors as well as structural ones.

The factor of safety is calculated by dividing the load required to cause failure by the maximum load expected to act on a structure. Thus if a rope with a capacity of 6,000 pounds is used in a hoist to lift no more than 1,000 pounds at a time, then the factor of safety is $6,000 \div 1,000 = 6$. As used in the design of a hoist, a factor of safety of 6 would be determined by experience or judgment. It would establish the size of rope necessary to allow for a margin of error and uncertainty in the materials and use of the hoist. For example, even though the hoist is rated at 1,000 pounds, operators can be expected to overload it, perhaps by half as much again as its capacity. Thus a 1,500-pound load could be on the hoist at any give time. Furthermore, even though the operator might have instructions to start and stop the hoist slowly, it could be expected to be operated in a jerky manner at times, causing the effective load on the rope to be increased due to the effects of inertia. If this could as much as double the effect of the 1,500-pound load, then a load of 3,000 pounds might in fact be applied to the rope.

Not only might the load be as much as three times that intended by the rope designer, but also the rope itself might be weaker than the specified 6,000 pounds for several reasons. An inferior rope may have been unwittingly installed or it may have become frayed and weakened due to use or abuse. If the net effect were that the rope could barely support 3,000 pounds, we can see that the jerky lifting of an overloaded hoist bucket could break the rope and the hoist would fail. The factor of safety of 6 would not have been enough to have anticipated all the adverse conditions occurring simultaneously. If there were no limit as to how badly overloaded or how severely jerked the hoist might be or how damaged or inferior the rope might become, then there would be no reasonably small number to choose as a factor of safety. To obviate failure of the hoist, an engineer might have to make the rope hundreds of times stronger than the rated capacity of the hoist. This would clearly be an uneconomical and bulky design and one that might easily be undersold by less conservative competitors.

Fortunately, by direct and indirect means, limits can be placed on how much of an overload might actually be expected to be applied to the rope of a hoist. The motor of the hoist can be made only strong enough to lift, say, 1,400 pounds, and it can be designed to accelerate and stop only with a relatively jerk-free motion. The rope can be proof-tested, or "proved," by hanging, say, 3,000 pounds from it before it is installed, but it cannot be proof-tested to its full expected capacity because that alone might destroy the rope. As the hoist ages, careful inspections of the rope for frayed or cut strands can be called for at regular intervals. All of these kinds of precautions can give the engineer designing the hoist the assurance that a factor of safety of 6 is not too low.

The essential idea behind a factor of safety is that a means of failure must be made explicit, and the load to cause that failure must be calculable or determinable by experiment. This clearly indicates that it is *failure* that the engineer is trying to avoid in his design, and that is why failures of real structures are so interesting to engineers. For even structures that fail are designed with various factors of safety, and clearly something went wrong in the engineering reasoning, in the construction, or in the use of the structure that fails. By understanding what went wrong, any misconceptions in the behavior of materials or structures can be corrected before the same mistake is made again. Generally speaking, when structural failures occur, a larger factor of safety is used in subsequent structures of a similar kind. Conversely, when groups of structures become very familiar and do not suffer unexplained failures, there is a tendency to believe that those structures are overdesigned, that is to say, they have associated with them an unnecessarily high factor of safety. Confidence mounts among designers that there is no need for such a high factor of ignorance in structures they feel they know so well, and a consensus develops among designers and code writers that the factor of safety for similar designs should in the future be lowered. The dynamics of raising the factor of safety in the wake of accidents and lowering it in the absence of accidents clearly can lead to cyclic occurrences of structural failures. Indeed, such a cyclic behavior in the development of suspension bridges was noted following the failure of the Tacoma Narrows Bridge.

The factor of safety is not a new concept, and in 1849 the Royal Commission appointed to investigate the use of iron in railway bridges asked of the prominent engineers of the time, "What multiple of the greatest load do you consider the breaking weight of the girder ought to be?" The answers from the likes of Robert Stephenson, designer of the Britannia Bridge, Isambard Kingdom Brunnel, the engineer of the Great Western Railway, and Charles Fox, engineer of the Crystal Palace, ranged from 3 to 7. And when asked, "With what multiple of the greatest load do you prove a girder?" the panel responded with factors ranging from 1 to 3. The commission concluded that an appropriate factor of safety for railway bridge girders would be 6.

Since a real structure will have not only beams and girders but also columns and other structural elements, there can be many ways that it can fail and thus many factors of safety associated with the structure. Generally it will be the smallest factor that is spoken of as *the* factor of safety of the structure. Structural studies done by Professor David Billington at Princeton University calculated the factors of safety for the Washington Monument by considering three possible ways in which it could fail: by crushing of the stone at the base due to the massive weight of the stone above it in the obelisk; by overturning in the wind; and by cracking under the action of the wind. The first two imagined failure modes were found each to have associated with them a factor of safety of approximately 9, while the cracking mode yielded a factor of safety of only 3.5. Since the force due to the wind is proportional to the square of the wind speed, this factor of safety means that a wind speed almost twice as high as any ever expected in the District of Columbia would have to blow to topple the famous monument. Hence its factor of safety of 3.5 would seem to be adequate insurance against its failure by any reasonably conceivable means.

While a factor of safety should be present implicitly in all engineering design judgments, sometimes the factor is used quite explicitly in calculations. Somehow neither seems to have been done or done correctly in the design of the Hyatt Regency walkways. The post-accident analysis of the skywalk connections by the National Bureau of Standards determined that the originally designed connections could only support on the average a load of 18,600 pounds, which was very nearly the portion of the dead weight of the structure itself that would have to be supported by each connection. Hence the factor of safety was essentially 1— which leaves no margin for error and no excess capacity for people walking, running, jumping, or dancing on the walkways. How a design with so low a factor of safety came to be built is a matter before the courts, and because of the numerous suits and counter-suits, the full story may not yet have been told. However, one may speculate as to what might happen should one wish to span an atrium without obstructing floor traffic with columns.

To design such walkways as those in the Kansas City Hyatt Regency means first to have a general idea of how to span the 120-foot space over the hotel lobby. If this was to be done without obstructions on the floor, then we can imagine how the skywalk concept arose. The idea was to get hotel patrons from one side of the lobby, where their rooms were, to the other side of the lobby, where meeting rooms and a swimming pool were located, without everyone having to go down to the ground floor, walk across the (crowded) lobby, and then go up to their destinations. The functional and architectural requirements suggest the idea of a bridge, with the two sides of the lobby as banks and the lobby floor a river or harbor whose traffic is to be unobstructed. These are exactly the requirements that suspension bridges must meet, and thus to suspend walkways from the ceiling is no great leap of the imagination. Since the lobby was four stories high, there was need for three separate bridge levels, and somehow the extrastructural decision was made to place the skybridge from the fourth floor directly above that from the second floor and to offset the skybridge from the third floor. With this general layout in mind, the designer's task is to select the structural parts by which to effect the concept. While the spacing of roof beams may in some way determine how close or how far apart suspender connections may be placed, there are still plenty of decisions to be made regarding the size of steel rod, the style and size of beams, and the details of connecting them all together to achieve the desired effect with a reasonable factor of safety.

Sizing the different parts can be tricky, for the size of beams and girders under the walkway determines the weight of the walkway, and the weight of the walkway in turn determines the sizes of beams and rods required to support it all. The smallest possible beams will be the lightest and thus lead to a lighter and less expensive structure. But if the beams are too small relative to their length they not only might be too weak to resist being broken or bent under their own weight and the weight of the concrete floor and the people on the walkway, but also might be too flexible and thus bounce too readily under the feet of running children or bend to a curvature too pronounced to be in harmony with the straight architectural lines of the atrium.

The designer can proceed to choose parts, assemble them on paper, and then calculate the various loads and deflections and factors of safety that he identifies to be important. It is here that the designer is really considering how the structure can fail—either by sagging too much or by placing too much load on some individual beam, rod, or connection—for it is only by having an idea of failure that he can calculate a factor of safety. Experience of prior structural successes and failures can be of immense help at this stage of design, for what has worked under analogous structural requirements enables the designer to size and detail his structure with some degree of confidence, while a knowledge of what has failed to work alerts him to pay special attention to what are potentially weak links. The Monday-morning quarterbacking of seasoned designers and detailers in response to the design and details of the ill-fated skywalks seems to have been unanimously critical of the choice of the flimsy box beam and hanger rod connections, but evidently the original designer and whoever made the change either did not recognize this detail as a potential weak link or miscalculated its actual factor of safety.

The example of the poorly conceived skywalks emphasizes the view of design as the obviation of failure. While the initial structural requirement of bridging a space may be seen as a positive goal to be reached through induction, the success of a designer's paper plan, which amounts to a hypothesis, can never be proven by deduction. The goal of the designer is rather to recognize any counterexamples to a structurally inadequate hypothesis that he makes. In the case of the atrium skybridges, the hypothesis was that the design as built would span the space above the lobby without falling. The truth of this main hypothesis could never have been proven, but its falseness could have been established by analyzing the rod-box beam connections and finding that they could indeed fail to perform under the expected loads.

Each novel structural concept—be it a skywalk over a hotel lobby, a suspension bridge over a river, or a jumbo jet capable of flying across the oceans—is an hypothesis to be tested first on paper and possibly in the laboratory, but ultimately to be justified by its performance of its function without failure. However, as the examples of the Kansas City hotel skywalks, the Point Pleasant Bridge, and the DC–10 airliner demonstrate, even success for a year or years after completion does not prove the hypothesis to be valid. Yet were we not willing to try the untried, we would have no exciting new uses of architectural space, we would be forced to take ferries across many a river, and we would have no trans-Atlantic jet service. While the curse of human nature appears to be to make mistakes, its determination appears to be to succeed.

Technology has advanced by our constantly seeking to understand the hows and whys of our own disappointments, and we have always sought to learn from our mistakes lest they be repeated. But failures do and will occur because new structural designs or materials are continually being introduced into new environments, and there is little indication that innovation will ever be abandoned completely for the sake of absolute predictability. That would not seem to be compatible with the technological drive of *Homo faber* to build to ever greater heights and to bridge ever greater distances, even if only because they are there to be reached or spanned. Each new structural hypothesis is open to disproof by counterexample, and the rational designer will respond immediately to the credible failure brought to his attention.

Although the design engineer does learn from experience, each truly new design necessarily involves an element of uncertainty. The engineer will always know more what not to do than what to do. In this way the designer's job is one of prescience as much as one of experience. Engineers increase their ability to predict the behavior of their untried designs by understanding the engineering successes and failures of history. The failures are especially instructive because they give clues to what has and can go wrong with the next design—they provide counterexamples.

Most engineering design hypotheses that are constructed do not fail, of course, but the structural success of another traditional design is no more news than the man who does not rob a bank or does not bite a dog. It is the anomaly that gets the press, and the abnormal that becomes the norm of conversation. Thus, to speak of engineering failures is indirectly to celebrate the overwhelming numbers of successes.

ON THE BRIDGE!

I. Failures of engineering have often been the subject of public outrage and ridicule. As nineteenth-century iron railroad bridges were developing cracks and failing at unacceptable rates, John Tenniel's macabre engraving of "Insatiable Death Awaiting Inevitable Accident" was published in *Punch* in 1891. And in 1979, when colossal engineering blunders seemed to be happening everywhere, editorial cartoonist Tony Auth offered his solution to the problem.

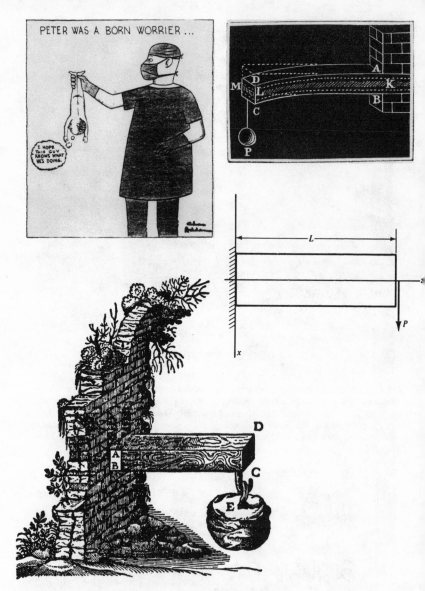

II. Constantly worrying about the possibility of being dropped—by a doctor's arm, an airplane's wing, or any of the many other cantilever beams we must rely upon—is no way to go through life. The object of engineering design is to anticipate failure and to design against it. This is done by understanding how much load a structure can carry without letting go or breaking. Galileo was the first to approach in a rational way the problem of a cantilever beam. His seventeenth-century analysis of the problem was less correct than the detailed woodcut of it, but as the engineer's understanding of the strength of materials became more mathematical and precise, the illustrations of the physical problem became more abstract.

III. The Brooklyn Bridge is one of the great engineering achievements of the nineteenth century, and its success depended in part on the lessons its designer, John Roebling, learned from the many contemporary suspension bridge failures. Here the bridge is shown under construction, with the temporary catwalk in place for use in spinning the cables. The cautionary sign installed at the insistence of Roebling's son, Washington, indicates the engineer's understanding of the sensitivity of long, slender, and flexible structures to dynamic loading. The roadway of the Brooklyn Bridge itself would have deep girders to give it the stiffness and strength to carry railway cars full of commuters between Manhattan and Brooklyn. In spite of the extremely stiff and safe design, only a week after the bridge opened in 1883 a crowd of twenty thousand panicked when rumors of collapse spread through the throng.

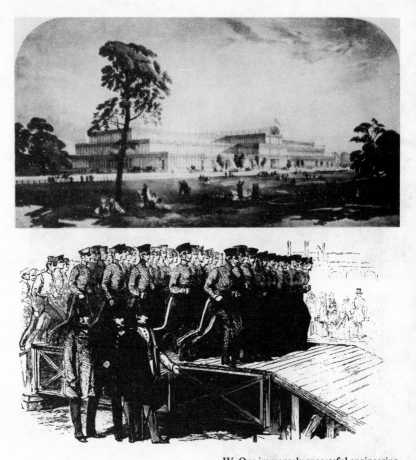

IV. One immensely successful engineering structure of the nineteenth century was constructed in 1850–1851 for the first International Exhibition and World's Fair. This enormous building covered almost nineteen acres of ground in London's Hyde Park and was completed in only seventeen weeks. The success and efficiency of the project was a result of innovative construction management techniques, modular design, and the engineering experience gained in building the early railroads. When the strength of the unique iron and glass structure was questioned, tests on a section of the building's galleries were conducted before Queen Victoria and her entourage. The success of the tests was celebrated in a Tenniel drawing in *Punch,* which had wholly supported the enterprise and had christened the great building the Crystal Palace.

New York Exposition & Convention Center
Rendered Interior View of Great Hall from Entrance

V. The Crystal Palace was over 1,850 feet long and 400 feet wide, and its influence is present today in the facade of the new Infomart in Dallas, Texas. This 1.5-million-square-foot facility is a marketing center for computers and related services, and the architect was so inspired by the original Crystal Palace that he endeavored to be faithful to its design, even to the Victorian color scheme. It was not so much the exterior but the vast interior space of the nineteenth-century building that inspired the design of the New York Convention Center, now under construction near the Hudson River. However, this latter structure has been plagued by delays and cost overruns uncharacteristic of the original Crystal Palace, but not of its reconstruction south of London.

VI. The failures of nineteenth-century suspension bridges were forgotten in the great successes of the Brooklyn Bridge and its structural descendants. As bridge designers strove for longer and lighter bridges, their confidence was upset by the spectacular failure of the Tacoma Narrows Bridge, which was twisted apart in a moderate wind in 1940. Sister bridges were immediately stiffened and new bridges were built with very deep and stiff roadways, such as that of the Mackinac Bridge, as seen here. In England, designers made the roadway of the Severn Bridge in the shape of a wing to slice through winds that might otherwise be its undoing. However, the bridge, shown here under construction in 1966, has developed severe cracking and fatigue problems, which have been attributed to its extreme lightness relative to the heavier traffic it has been allowed to carry since its design and completion.

As Built   Original Detail

VII. The collapse of two suspended walkways in the Kansas City Hyatt Regency Hotel in 1981 killed over one hundred people and was the worst accident due to a structural failure in the history of the United States. There were many theories offered for the cause of the collapse, including the presence of too many people dancing on the walkways, recalling the collapse of early suspension bridges under the feet of marching soldiers. The real cause of the failure was quickly traced to a single change in the design of a support detail, apparently made to facilitate the erection of the skywalks. *The Kansas City Star,* which had hired an engineer as a consultant on the story, revealed the true cause within days of the accident in this Pulitzer Prize-winning story. Later tests at the National Bureau of Standards confirmed the cause to be in the detail.

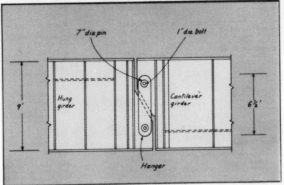

7" dia. pin   1" dia. bolt

Hung girder   Cantilever girder

9'   6½'

Hanger

VIII. A section of an Interstate 95 bridge across the Mianus River in Connecticut collapsed suddenly in the early morning hours of June 28, 1983. Three persons were killed in vehicles that fell through the hole, but there might have been many more casualties had the structural failure occurred during rush hour, for over 100,000 vehicles used this bridge daily. The collapse called attention to the pin-and-hanger design of the twenty-five-year-old bridge, which left the structure with no support when something went wrong with the pin. Subsequent inspections uncovered many cracked hangers in bridges of similar design. While the bridge designer blamed the State of Connecticut for not properly maintaining the bridge, the state blamed the designer for an inferior design. The accident itself called attention to the importance of *both* proper maintenance and good design.

10
WHEN CRACKS
BECOME BREAKTHROUGHS

The Liberty Bell, whose crack represents one of the most famous engineering failures in our history, symbolizes more than political independence. It also represents both the triumph of and the toll taken by the technological independence that was won as a result of the Industrial Revolution. The freedom to design and realize larger, more ambitious, and more complex engineering structures has never been without risk, however, and the early history of steam power and railroads is punctuated with the debris and casualties of boiler explosions and railway accidents. If there are still structural failures today, it is largely because there are still technological frontiers in the new industrial world. These frontiers are not crossed without risk, but neither are they crossed without precaution by responsible technological pioneers.

The history of the Liberty Bell itself forebodes the vicissitudes of technological development. The original bell arrived from England, where it had been made, in 1752, and it cracked the very first time it was rung. Instead of being sent on the long journey back across the ocean to be recast, the bell was melted down by local workmen in Philadelphia. After making some small bells to test the sound and strength of their castings, the inexperienced Americans decided to increase the copper content of the bronze. When the resulting large casting was found to have poor tone it was remelted. A small amount of silver was then added to sweeten the tone, but the Pennsylvania Provincial Assembly was not completely satisfied with the new casting, and there was some talk of getting a new bell from England. However, the colonial bell was finally found to be acceptable, perhaps owing to provincial pride, and it was rung on July 4, 1776, to signal the adoption of the Declaration of Independence.

The Liberty Bell was taken away in 1777 to be hidden from the advancing British army, and it apparently suffered some abuse in transit before being returned to Philadelphia the following year. For more than half a century, anniversaries and special events were marked by ringing the bell. Then suddenly, in 1835, while it was tolling for the funeral of Chief Justice John Marshall, a large crack appeared in the Liberty Bell. Attempts were made to keep the crack from spreading, but it extended to its now familiar length while the bell was being rung again in 1846 to celebrate George Washington's birthday.

Not all failures are spontaneous like the Liberty Bell's. Often structural failures develop slowly enough to be noticed before causing irreparable damage, and corrective measures can be taken to maintain the cracked structure's serviceability. This was the case with Big Ben, the thirteen-ton bell hanging in the east tower of London's Houses of Parliament. When a crack was discovered in the great bell over one hundred years ago, the clapper was replaced with a smaller one, which was also made to strike the bell in a less damaging spot. This corrective engineering action has succeeded in keeping Big Ben ringing the quarter hours (though some say with less pure a tone than it had) rather than being just a mute curiosity. The crack is still there, but it is not growing at any significant rate, nor is it expected to do so.

Big Ben's chiming mechanism also developed a crack, but it went undetected until the flawed shaft of the fly governor broke catastrophically in 1976 and caused extensive damage to the mechanism in the clockroom as clock parts flew about violently.

The crack in the shaft, initiated by a manufacturing flaw, had apparently grown slowly but deliberately over the four million striking cycles since the clock was installed in 1859—a classic fatigue failure.

Fifty to ninety percent of all structural failures, including those of bells, bridges, airplanes, and other commonplace products of technology, are believed to be the result of crack growth. In most cases the cracks grow slowly. It is only when the cracks reach intolerable proportions for the structure and still go undetected that catastrophe can occur. Thus cracks per se need not be cause for alarm, and responsible engineering design takes into account the possibility that cracks or other flaws of material or worksmanship will be present in the object designed. The effects of those flaws on the structure throughout the course of its lifetime can be calculated as part of the design, and engineers can alert owners and operators of their structures to be vigilant for growing cracks lest the calculations themselves be flawed.

"Brittle fracture," in which a large crack runs spontaneously through a structure near the speed of sound, severing steel with a report that signals the breakup of ships, the bursting of pressure vessels, or the collapse of bridges, has been a chronic problem for centuries. There is almost always a period of "gestation"—the slow lengthening and sharpening of cracks through the process of fatigue that precedes the catastrophe. Carl Osgood, in his monograph, *Fatigue Design,* even goes so far as to assert: "All machine and structural designs are problems in fatigue because the forces of Nature are always at work and every object must respond in some fashion." And he reiterates the assertion in the preface to the second edition of his book, having in the intervening decade "found no reason to alter [his] original statement." While not all engineers are so exclusive in their concern with fatigue, it is certainly an important consideration in many, many designs.

There are several metallurgical theories to explain the mechanism of progressive fatigue damage, including elaborate hypothesized irregularities or "dislocations" in the metal microstructure, but none is wholly satisfying. Yet even as metallurgists debate the microscopic details of exactly how a piece of metal breaks, engineers are constantly being called upon to design machines and structures that will not break even when subjected to extreme vibrations and other varying loads. Thus engineers must develop practical methods for predicting how fast cracks grow and how large they can grow without causing failure. It is often such considerations that establish a structure's projected useful lifetime.

If we consider that our understanding of the process itself is incomplete, the success of structural engineers in designing against failure by fatigue seems admirable. For several decades they have viewed the fatigue process as consisting of essentially two stages. Microscopic cracks develop at "nucleation sites"— points of material weakness or stress concentration—during the first stage, which may occupy as much as one-half of the entire lifetime of a machine part or structure. As repeated loading continues, these cracks grow and some coalesce into a dominant macroscopic fatigue crack. During the second stage, this crack grows at an accelerating rate as the cycles of loading continue. If the crack reaches an intolerable size for the load being applied, the weakened structure can no longer support it. The crack makes a final advance under a load level that may be well within the design load of the unfatigued structure.

Metallurgists have learned, often empirically, to produce alloy metals with as few nucleation sites and as much resistance to crack growth as possible. And engineers have learned to beef up joints to reduce localized load levels and to use materials with large capacities for containing cracks without suffering brittle fracture. But the problem of metal fatigue persists because metallurgist and engineer alike attempt to predict, from limited past experience, the behavior of ever-new materials in an uncertain future environment of use and abuse. The slightest deviation from experience in a new design can have unanticipated consequences. And in some cases the crack gestation stage is so short as to be virtually absent, as in the case of metallurgical disasters like the Liberty Bell's.

Understanding the phenomenon of fatigue and preventing it are two entirely different things. Hypotheses about crack growth are tested under the ideal, controlled conditions of the laboratory. Test specimens can be carefully machined to provide as flawless a surface as possible, and loading conditions can be carefully specified and monitored. Since replication of test results is achievable under such conditions, smooth curves representing repeated load levels or stress (often designated by the letter S by engineers) versus number of cycles to failure by fatigue (designated N) are easily generated. These S-N curves characterize the behavior of each type of material. Of course, as the stress is diminished, the number of cycles to failure—the structure's "lifetime"—increases. Furthermore, if the load level is diminished below a threshold value, failure is *never* observed no matter how many cycles of loading are applied.

Fatigue can thus be theoretically avoided, but overdesigning structures so that peak stresses never exceed the threshold level is not practical. An airplane designed that way might be too heavy to fly. And even if it could, a competing manufacturer would soon produce a lighter design that was less expensive to build, buy, and operate. Yet "optimal" design, in which fatigue-crack propagation is acknowledged to occur, but so slowly that the structure will have long been taken out of service before the crack poses any safety problem, is also more easily conceived in principle than achieved in practice.

Although the simple S-N curve is readily available for most engineering materials, its use in design is not so straightforward. For one thing, the curves must be interpreted statistically, with ranges of uncertainty being as important as points on smooth curves. Not only will there be scatter in the collection of fatigue data, there is variation among batches of the same material and even parts of the same batch used in different physical and chemical environments.

Furthermore, fabricated structures are not nearly so flawless and free of crack nucleation sites as their laboratory models. Carelessness such as that which leads to faulty welding not only can introduce important nucleation sites but also can provide massive fissures in the metal that bypass the crack nucleation phase altogether. The fabrication stage also sometimes introduces unknown residual stresses in the structure, and these can be the principal driving forces of the early stages of the fatigue process.

Finally, the actual loadings to which engineering structures are subjected are not nearly as regular or mathematically precise as test loads employed under laboratory conditions, and sometimes the actual conditions are not even taken into account by the designer because he does not imagine them to be credible. This was the case with the DC-10 engine mount, in which large cracks were introduced through severe and unanticipated shortcut maintenance procedures.

"Quality control" is supposed to eliminate unacceptably large flaws by minimizing deviations from an acceptable norm and by rejecting inferior workmanship. But, unfortunately, the techniques for the detection of preexisting cracks in fabricated structures are wanting. Not only are instruments relatively insensitive, but also their use and interpretation are often more art than science.

One of the most promising tools of "nondestructive testing"—the collective term for techniques used to probe for flaws within opaque parts and the welds that join them without destroying the object—employs ultrasonic waves. By sending these waves into a part and observing their return, the internal integrity of the otherwise invisible interior of the object can be determined without the structure or any part of it being broken in the process of looking. A fissure, void, foreign body, or other potentially troublesome flaw in the metal will reflect ultrasonic waves in a characteristic way to a receiving transducer. Unfortunately, the mathematical characteristics of reflected waves are incompletely known for most of the complex geometries that comprise modern structures. And the interpretation of signals on an oscilloscope filled with noise and interference from other parts of the structure can leave a lot to the imagination.

Other nondestructive testing techniques based on well-understood physical phenomena such as X-rays and magnetic fields simply do not have the resolution and sensitivity to give sufficiently detailed information about an engineering structure. Negative results from the tests simply mean that no *large* flaws were detected by the equipment and operators. In some cases, cracks extending as much as a quarter of the way through a steel plate can escape notice in nondestructive testing procedures.

A variation on nondestructive testing is the "proof test," in which a newly completed pressure vessel or other container is slowly pressurized or a beam loaded beyond what it is expected to withstand under the worst service conditions. Of course, this kind of test is nondestructive only if the vessel or structure proves sound. Such a technique would seem to demonstrate conclusively the integrity of the part, but this type of test, too, is limited; flaws small enough not to have caused the vessel to rupture during the test but large enough to cause failure after a significant period of service and fatigue-crack growth could well be present.

The existence of undetected small flaws can be hypothesized from the observed strength of the structure or inferred from the sensitivity of the testing technique. Engineers must assume that such cracks exist and calculate their rate of growth. For example, when an airplane is constructed, quality-control procedures, proof tests, and test flights ensure that no immediately dangerous large cracks exist in the structural members. Yet initially benign cracks may indeed exist, and engineers must determine how long it should take such cracks to reach a detectable size and thus how soon an in-service inspection should be made. For further protection from catastrophic failure owing to growth of a fatigue crack to critical proportions, engineers usually employ "fail-safe" designs: those that incorporate structural obstacles to the spontaneous growth of cracks that might escape detection.

A second basic design philosophy to obviate structural failure is called the "safe-life" criterion. Safe-life design, which allows for the inevitability of failure well beyond the service life of the structure, is not so simple to realize. The engineer can ascertain, relatively easily in many cases, what might be the worst initial flaw in a structure, but specification of exactly how the structure will subsequently be loaded and exactly how the crack may grow is not possible. For example, a design engineer can estimate the routine loads on an airplane during its lifetime of takeoffs, climbs, flights, encounters with turbulence, landings, and taxiings, but he cannot easily predict the exact progress of a single fatigue crack throughout such cycles. That depends upon the exact sequence of flights during storms and calm weather, soft and hard landings, and light and heavy payloads. Moreover, the mathematical models for analyzing the effects of different load levels in different orders of occurrence are far from standard. Therefore, a safe design must not rely entirely upon the safe-life criterion but must also be fail-safe so that potentially dangerous cracks will be arrested and caught by regularly prescribed inspections.

Despite the limitations of analysis, whether from inadequate theories or uncertain input, design engineers still have to do their jobs. And since real-time, full-scale fatigue experiments are generally out of the question—to build a full-size model of the structure is to build the structure itself—engineers use factors of safety and incorporate fault-forgiving features into their designs.

Whether cracks were growing to dangerous sizes in the walls of nuclear reactor vessels became a hotly debated issue during the mid-1960s among members of the nuclear industry, the Atomic Energy Commission, and that agency's Advisory Committee on Reactor Safeguards. ACRS members recognized that it was impossible to guarantee that steel reactor vessels with six- to eight-inch-thick walls were flawlessly manufactured. Because fabrication and inspection techniques were less than perfect, one simply had to assume that defects of minimal size existed, but what constituted a defect of minimal size is a critical assumption that is still subject to some debate.

Furthermore, over the decades during which nuclear plants are expected to operate, cycles of heating and cooling, fluctuations in power and pressure levels, and general variations in the stresses in vessel walls could cause fatigue cracks to develop and grow. Whether these cracks could be detected was as uncertain then as it is now given the difficulties of inspecting nuclear plants while in operation and given the limitations of available instruments. Nondestructive testing techniques, such as using remote-controlled television cameras that are moved about the vessel to aid inspectors looking for cracks and other suspicious breaches of the metal's integrity, are severely limited in their reliability. Therefore, certain types of cracks must be assumed to exist, even though analysis via the techniques of fracture mechanics could reasonably demonstrate that the defects were benign.

But the issue is complicated in nuclear reactor plants by the fact that the steels from which pressure vessels are made do not have a single "fracture toughness"; their resistance to catastrophic fracture depends upon the temperature of the metal. There is a certain temperature below which the material is very brittle and not tough at all. This is called the "reference temperature" when it incorporates a safety margin equivalent to a factor of safety. As long as the reactor vessel is pressurized well above this temperature, there is no danger of any crack opening catastrophically.

But when steel is cooled below the reference temperature, brittle fracture can occur. This phenomenon was partly responsible for the sudden collapse in winter of some early welded bridge designs and for the spectacular breakup of Liberty ships during World War II. A typical ship failure occurred on January 16,

1943, in Portland, Oregon. The weather was mild and the sea was calm. A T–2 tanker was lying quietly at pier when, without warning, she suddenly broke in two with a report that was heard for at least a mile. Other spectacular breakups of tankers and Liberty ships, mostly fabricated using the newly adopted all-welded construction techniques, followed over the next several years. (Ironically, the welded seams replaced the traditional riveted ones that automatically stopped any cracks from spreading from plate to plate.) Thus the phenomenon of brittle fracture, which had been occurring at least since the use of steel in bridges, storage tanks, and other engineering structures, came to be studied more and more, and the engineering science of fracture mechanics came into existence approximately at the same time as nuclear reactor vessels. Research in the wake of the failure of Liberty ships determined that spontaneous fracture occurs under critical combinations of metallurgy, environment, and load, usually aggravated by the presence of geometrical irregularities such as sharp corners or cracks. In the case of the welded ships, hastily implemented techniques and inexperienced workers further threatened the integrity of the steel structures. Research has taught engineers how to design against brittle fracture, but the challenge remains to establish the size of preexisting flaws in structures and to predict the growth of those flaws into fatigue cracks.

Nuclear reactor vessels are designed to operate well above the reference temperature of their steels. Thus any cracks that might be undetected by remote-control television or other means of inspection would not be dangerous because the vessel would not be brittle like a welded ship in the cool sea or a hot glass coffee pot on a cold tile counter. Unfortunately, one of the effects of neutron irradiation from the reactor's core is to raise the reference temperature over a plant's lifetime. This fact was known in the mid-1960s, but there was no experience then to indicate exactly how high the critical temperature of today's reactors would be raised. Engineers and metallurgists made what they considered conservative projections of the damaging effects of the neutron flux and designed their reactors accordingly. In the early 1980s, however, only fifteen years later and only about halfway through the lifetime of most early reactors, tests on irradiated steel suggested that the embrittling process had progressed much faster than anticipated. When first installed, these reactor vessels could be operated as low as 100° Fahrenheit and still remain well above the reference temperature. Some reference temperatures were estimated to have changed with the steel's irradiation to over 200° Fahrenheit, and although reactors normally operate at about 600° Fahrenheit, the perilous combination of a lower temperature and a high pressure during an accident could not be completely ruled out. Thus, should an accident occur that would trigger the emergency core cooling systems to cool the pressurized vessel too quickly—toward a raised reference temperature—any preexisting cracks might grow quite rapidly.

Power companies who operate nuclear reactors in the United States closely monitor the fracture toughness of their vessels, and these operators generally believe that embrittlement has not progressed to dangerous levels. Steadily improving techniques for detecting cracks and analyzing cracks also make the operators confident that they are taking no unnecessary risks. Furthermore, reactor operators expect that annealing techniques being developed will soften and toughen the steel to regain lower reference temperatures, thus reversing the embrittling effects of neutron irradiation and restoring the safety margin to earlier levels. The British, however, have continued to debate the issue of possible dangerous cracks in American-designed pressurized water reactor vessels that have not yet been used to generate electric power in their country.

The nuclear power industry has also had its share of problems with cracks in pipes, and the potential for a nuclear plant accident harming thousands of people, as a film like *The China Syndrome* dramatizes, has threatened the shutdown of nuclear reactors even at the risk of causing power shortages. Some of the most troublesome kinds of cracks in nuclear plant piping are related to corrosion, and they were thought to have been obviated by the careful choice of stainless steel for the pipes, the careful manufacturing and installation of the pipes, and the careful control of the chemistry of the water that flows through them. Reactor cooling water, in particular, was kept unusually pure. But the untried combinations of stresses, welding techniques, temperatures, and water chemistry were invitations to surprise. Thus the discovery during the late 1960s and the early 1970s of several occurrences of cracking in the pipes in nuclear plants eventually led to the establishment, in early 1975, of a Pipe Crack Study Group by the then newly formed Nuclear Regulatory Commission.

The study group attributed the cracks to what is known as intergranular stress-corrosion cracking, a condition that occurs in susceptible metals under an undesirable combination of relatively high mechanical stress and a corrosive environment. The condition is characterized by cracks proceeding generally along metallurgical grain boundaries, which might be thought of as microscopic fault lines, and is enhanced by the presence of crevices where corrosive agents collect and mechanical stresses are raised. Metals can be made susceptible to stress-corrosion cracking during the welding process, and the high oxygen content in the purified nuclear plant cooling water can aggravate the situation. Some stories, perhaps apocryphal, contend that pipes are made susceptible when welders heat one side of a pipe section to bend it into alignment with another pipe section to which it must be welded. Such an unauthorized and unanticipated procedure would introduce stresses no design engineer had factored into his calculations. Another story, again perhaps apocryphal, has welders at lunchtime straddling the pipes they are working on and eating hard-boiled eggs. Some of the salt they sprinkle on the eggs falls, along with salty perspiration, into the pipe joints, and thus introduces sodium chloride into a system design engineers and operating engineers assume to be free of the known corrosive agent.

No matter how far-fetched the causes of cracks, a nuclear power plant must have an extremely reliable system of large-capacity pipes to supply coolant to the hot reactor core. A sudden rupture in one of the main supply pipes could lead to a major "loss-of-coolant" accident. Since cracks in pipes may not be detected during inspections or tests, and since crack propagation by fatigue cannot be ruled out, reactor designers have devised a safety concept known as the "leak-before-break" criterion against which to test their designs. If a certain type of ductile steel is used for the pipe wall, any crack that develops will grow faster through the wall of the pipe than in any other direction. This ensures that a crack will cause a relatively small but detectable leak well before a dangerously long crack can develop parallel to a weld seam and lead to a catastrophic rupture of the pipe. The criterion has been found dependable, and operating reactors have developed leaking cracks that spread across seventy-five percent of the pipe wall without leading to brittle fracture. When such cracks have been found by tracking down the leaks, the bad pipe section has been replaced with sound new piping. In spite of all that could and has happened to nuclear reactor piping, the Pipe Crack Study Group found the stainless steel used in the piping to be extremely tolerant of the largest cracks. The very conservative design of nuclear power plants has proven to have sufficient factors of safety built into the system, and the leak-before-break criterion appears to be a sound concept.

Nevertheless the sophisticated structure of modern engineering is, like the simple chain of antiquity, only as strong as its weakest link. Where that link is located is often an accident of metallurgy or luck, and today's engineer, like the historical blacksmith, would like to keep an eye out for and eliminate the potential trouble spot. Yet, sweltering at his forge on a hot summer day, the smith could no doubt lapse into daydreams and beat one link thinner than all the rest in his chain. Or he would attend to other matters while his apprentice overworked another link, perhaps cracking it. Or he would not cool an overheated link properly. Or the blacksmith would inadvertently skip a link when inspecting the finished product. Or he would not pull the chain to the fullest test load because one of his horses was ill. Or the chain's purchaser would attempt to get more than the blacksmith promised out of the chain. No matter how well today's engineer understands the behavior of cracks, he cannot factor in the unknown human element, including that of his own limitations of prescience, into all his calculations. The safe operation of a complex system like a wide-bodied aircraft or a nuclear power plant ultimately depends upon the strength and reliability of all the links, both human and mechanical, in the chain.

The consequences of structural failure in nuclear plants are so great that extraordinary redundancy and large safety margins are incorporated into the designs. At the other extreme, the frailty of such disposable structures as shoelaces and light bulbs, whose failure is of little consequence, is accepted as a reasonable trade-off for an inexpensive product. For most "in-between" parts or structures, the choices are not so obvious. No designers want their structures to fail, and no structure is deliberately underdesigned when safety is an issue. Yet designer, client, and user must inevitably confront the unpleasant questions of "How much redundancy is enough?" and "What cost is too great?"

When a seemingly common structure such as a bus frame is required to support a new lightweight, energy-saving vehicle, or an elevated walkway is specified to span an architectural space without obtrusive columns, designers confidently accept the challenge. They know that engineers' experience with virtually indestructible traditional bus frames and bridges spanning a mile make the achievement well within their capabilities. And new and sophisticated analytical techniques, modern high-strength materials, and the aid of computers provide even greater confidence. But these advances can also make engineers overconfident that they can depart dramatically, and perhaps prematurely, from traditional designs.

Thus sometimes mistakes occur. Then it is failure analysis—the discipline that seeks to reassemble the whole into something greater than the sum of its broken parts—that provides the engineer with caveats for future designs. Ironically, structural failure and not success improves the safety of later generations of a design. It is difficult to say whether a century-old bridge was overdesigned or how much lighter the frames of forty-year-old buses could have been.

11

OF BUS FRAMES
AND KNIFE BLADES

When fifteen thousand runners swarm across the Verrazano Narrows Bridge to begin each annual New York City Marathon, no one seems to worry about the rise and fall of all those running shoes coming anywhere near exciting the roadway to resonance the way marching soldiers did to many a nineteenth-century suspension bridge. And the collapse of sky-walks under dancing crowds was probably also far from the minds of the marathoners in 1981 as they ran off the bridge into Brooklyn, proceeding along Fourth Avenue toward Flatbush. If they were thinking of anything, it might have been of pacing themselves to finish the race without an ounce of strength to spare, to run with virtually no factor of safety, to go more than twenty-six miles and finish just as they are about to drop from fatigue, to endure several hours the way the deacon's one-hoss shay lasted one hundred years.

Among the lead runners in the twelfth New York Marathon was Alberto Salazar, who had collapsed in an earlier race. However, he had learned from his failure and went on to cross the finish line in Central Park first this time. The runners were led by a television camera truck that broadcast live pictures of the race, and the multifarious undulations, bumps, cracks, and potholes in the New York streets were evident in the shaky pictures transmitted. It looked as if someone were running backward before the pack, shooting *cinéma vérité* with a hand-held camera. As the marathoners turned left off Lafayette onto Bedford Avenue, they were but four blocks from Quincy Street, where only months before a specially instrumented city bus was being put through its paces, not unlike the way a runner might be observed on a treadmill. Instead of cardio-vascular leads, the bus was loaded with sandbags and fitted with strain-gauges that told engineers how much bending and stretching was going on in the steel skeleton beneath the sleek skin of the relatively new bus, but the object was to determine the vehicle's resistance to metal fatigue, much as a team of doctors might wish to determine an astronaut's limits of endurance. The engineers studying the bus negotiating the potholes of Quincy Street wanted to know why its design was developing severe fatigue cracks before going less than one-tenth the distance it was designed to run. They were hoping to understand what went wrong so that the buses, like Alberto Salazar, could come back from defeat to gain victory over New York City's streets.

The buses that were not finishing their marathon service along the city's routes were known as Grumman Flxibles after their manufacturer, the Grumman Flxible Corporation, a division of the airplane company. (The name, which looks like a typographical error, was derived from the name for the flexible couplers that connected sidecars to motorcycles that the original Flxible Corporation made during World War I.) The unconventional designs of the buses were constrained by specifications of the Federal Urban Mass Transit Administration, which was subsidizing the purchase of the buses, and wanted models that would be lighter to save fuel, accept handicapped riders, be air-conditioned, and have electronic controls. Grumman developed a lightweight bus by radically changing the heavy chassis that had served so many earlier generations of virtually indestructible vehicles, and was the low bidder for what eventually was an 850-bus order by New York City. Other cities, such as Houston, Los Angeles, and Atlanta also were to purchase Flxibles for their fleets.

But New York withdrew its entire fleet of what was then 637 Flxibles from service after only about six months because so many fatigue cracks had developed in the lightweight frames. Continued use of the buses, it was argued, could have caused the weakened frames to break suddenly and lead to accidents. Some of the disabled buses were parked on a Hudson River pier while the City negotiated with the Grumman Corporation over what would be done. The company beefed up the inadequate frames, and the buses were returned to service about a year later. Grumman tried to hire Jackie Gleason, who played a bus driver on "The Honeymooners," to make commercials for an image-building campaign for the returning buses and the company. He reportedly wanted too much money, and Telly Savalas, whose Kojak character is believed to be a prototypical New Yorker, appeared in television, radio, and newspaper advertisements for the bus. The company also changed the name of the bus to "Metro," but the fatigue problem contributed greatly to Grumman Flxible's loss of over $30 million in the fourth quarter of 1981. This was a significant financial blow to the parent company, which was counting on the bus business to help it through the lean years of defense spending. Grumman eventually sold the bus subsidiary.

Trouble continued with New York's Flxibles, however, as cracks developed in the floor near the rear wheel wells, problems arose in the front axle assembly, and steering column bearings began to break up. In early 1984, the New York City Transit Authority (NYCTA) decided to get rid of all of its Grumman Model 870 Flxibles, which were supposed to last at least into the 1990s, and sued to recover the $92 million spent for the buses and for $350 million in punitive damages. The suit over the buses, plus the Navy's impending announcement of a contract to upgrade the Grumman-built A–6 and F–14 warplanes rather than the competitor McDonnell-Douglas F–18 aircraft, seems to have prompted a curious quarter-page advertisement on the op-ed page of *The New York Times* of July 23, 1984. The ad, with more white space than print, read in full:

> Other transit systems
> have driven
> their Grumman 870 buses
> enough miles to get
> to the moon and back 80 times.
>
> But the NYCTA claims
> it can't get their Grumman 870's crosstown.
> GRUMMAN

Grumman's ad may not be grammatically impeccable, but its message clearly suggests that New York's transit authority acted precipitously in giving up the bus. Whether or not the other transit systems operate routes so full of potholes is not made clear, however.

The company's protests notwithstanding, the Grumman Flxible bus will remain a classic engineering case study of design failure, even though it caused no catastrophic accident like the Tacoma Narrows Bridge or any deaths like the Hyatt Regency walkways did. The bus is a forceful example of what can go wrong when too many demands—fuel efficiency, light weight, accessibility, comfort, maneuverability, and more—require radical change, and especially reduction of strength, from what had worked successfully for so long. Who would have thought that latter-twentieth-century engineers could not design a metropolitan bus? It may be argued that New York City dramatized its troubles with the buses and could have lived with buses breaking down, for the threat to life was really quite remote. But one can also imagine life-threatening consequences if a bus were suddenly to drop an axle or have its chassis break in two or have its steering shaft bind or break. The probability of catastrophe is extremely difficult to calculate in a bus that seemed to develop one unexpected problem after another. And projections of maintenance costs would be difficult to make with confidence.

The discovery of cracks in metal can indeed be a perplexing problem, for their origin, growth, and threat can be difficult to assess, especially when there are questions about how the cracked structure was designed, manufactured, and used. Not long ago I had nine stainless steel knives arranged before me on my desk in a formation not unlike that of the Grumman Flxible buses that were parked on the New York City pier. Like the buses, a significant number of the knives had developed cracks, which were clearly visible in three of the nine knives at the base of the blade, near the handle. While the cracked knives *seemed* to be serviceable, the fissures made the metal look vulnerable and fragile, and I tried to ascertain their cause and possible effect on future use of the knives.

The possibility and consequences of steel letting go are no more difficult to imagine for knives than for buses. I could see myself struggling at dinner one evening to slice a piece of meat when the blade might encounter a bone. The crack would open suddenly and the inertia of the handle I would be holding could drive my fist down onto the edge of the plate. The whole thing could flip off the table, propelling the severed blade end over end through the air, out through my open window, across the driveway, into the open window of my neighbor, through the bars of his bird cage, and through the canary's throat. Who knows what else might happen? So while the knives sat arranged on my desk, my family got along with a few old, uncracked ones, and I conducted an informal safety analysis of the cracked blades, including a stress analysis of their design, a review of their manufacturing history and use, and a determination of their suitability for continued use.

The knives are from a set of stainless steel flatware that my wife used for a year or so before our marriage and that we had used continuously for fifteen years of everyday meals and informal entertaining. There had been no unfortunate accidents with the knives, though I might attribute this admirable safety record in part to our discovery about ten years earlier of the first cracked blade and the summary removal of that defective knife from our kitchen service. Subsequent discoveries of two less severely though still seriously cracked blades came when those knives could not be spared from our place settings. But the very rationality that enables man to make knives and buses drives him to reflect upon their imperfections. And when the talk at table went one evening from *The New York Times* reports of cracks in the then-new Grumman Flxible buses to the cracks in our old dinner knives, I could not enjoy further meals without thinking about the cracks in the knives.

The theoretical strength of metals that can be inferred from atomic bonds is not even approached in engineering structures. Conditions under which billions of diverse atoms are mixed at thousands of degrees and molded into tons of steel by men who sweat and blink and think of dinner are not those of the clean room of the physicist's mind. Metals, like men, are governed by statistics and probability, and only the probable can be counted on by the builders of bridges and buses and knives. The behavior of the ensemble in turn depends upon the behavior of individuals.

Imagining for the moment knives made of rubber and not of steel helps one to see the knife as an engineer might. The actions of slicing and spreading bend the blade against the object of the action, whether it is a tough or tender steak or a hard or flaky biscuit. The mechanical stresses accompanying this bending can be understood by imagining further that the rubber knife is covered with a coat of brittle paint. As children, many of us had such toy knives, and we remember how, as the knife was used, the paint cracked. This is because the material comprising the knife is stretched or compressed against its natural shape, and the paint cannot stretch or contract as much as the rubber beneath.

Stainless steel knife blades, though considerably stiffer than rubber models of them, do have, like rubber, a degree of elasticity and also spring back from their deformation. (I could see this clearly by holding a cracked knife before my desk lamp and flexing the blade. As I flexed it, the crack opened to let more light through, but it closed again like a camera shutter when I released my grip.) Just as the paint flakes off a rubber blade, so the surface of a metal one wants to flake or do something to relieve the tension of its position.

The most tenacious metal has a limit to the number of times it can be subjected to flexure and return to its pristine state. And the greater the degree of flexure, the fewer flexures it takes for damage to manifest itself. After so many cycles, even in the most carefully crafted piece, a minute crack will develop at a point of particularly high stress or susceptibility and will continue to grow until the piece is weakened to the point where it is not capable of enduring one more cycle of stress. Such cracks are fatigue cracks, and the endurance limit of the on-again-off-again stress of a mechanical or structural part is its fatigue life.

The cracks in our knives were all at the point of least metal and highest mechanical stress. Any predisposition to cracking could be expected to manifest itself in this location, and continued use of a flawed knife could propagate the crack through the steel. Though it may take millions or even billions of relatively small flexures of stiff steel to cause a crack to traverse a part, it will happen as surely as a paper clip will snap when bent back and forth a sufficient number of times.

Before we noticed the cracks in some of the blades, our knives were taken randomly from the silverware drawer, without regard to whether they were cracked or not. If our indiscriminate use spanned at the most, say, fifteen years of 365 days of three meals with an average of a dozen slicings and spreadings, that amounts at the very most to about 200,000 chances for a crack in a blade to open up under stress. By engineering standards, this is not a large number of repeated uses when it comes to growing cracks from scratch by fatigue. Furthermore, if our knives had been made equally and were used equally and suffered equally from purely mechanical fatigue, I would have expected them all to be cracked more or less equally after fifteen years. Yet they were not. (While three blades had cracks nearly halfway across the base of the blade, to this day six have no cracks at all.) This suggests that the cracked knives had a history different from the others. The cracks, though they might have grown minutely through fatigue, must have had their origins elsewhere than in purely mechanical means, and I looked more closely at the knives for clues to their differences.

The provenance of this stainless flatware is unknown to me beyond the identification

STAINLESS

JAPAN

on eight of the nine blades in this odd service. The ninth blade is distinguished by

STAINLESS

STEEL

JAPAN

This curious anomaly in an ostensibly matched set went unnoticed for fifteen years and came to light only when the knives were lined up like buses for inspection and reflection. Not only are the imprints not uniform in wording, but there also appear to be several different typefaces used in several different weights on the blades. We do not notice such details until we look for explanations, just as no one noticed the inadequate structural detail supporting the hotel walkways in Kansas City until they fell.

Prior to these discoveries I had assumed that the knives were virtually indistinguishable one from the other, though the selective cracking should have suggested from the start that there were unshared weaknesses and strengths. And when the knives were juxtaposed and not set separately among spoons and forks, it was apparent that the flatware was not reproduced by a single machine.

The design of the flatware is simple, and there are only a few bamboo leaves in bas-relief adorning the otherwise plain handles of the pieces. Scrutiny soon revealed that no two knives had exactly the same pattern of leaves, and when I grasped the set of knives to form a sheaf and stand it on end it was apparent that they were not all the same length. Thus the knives bear the unmistakable mark of handiwork: individuality.

The metal of these knives most likely started with a recipe for stainless steel, perhaps for the type whose solution of iron and carbon along with other ingredients is named after the English metallurgist of the Victorian era, Sir William Chandler Roberts-Austen, and which the American Iron and Steel Institute designates Type 304:

AUSTENITIC STAINLESS STEEL

1,400 lb. Iron
360 lb. Chromium
160 lb. Nickel
40 lb. Manganese
20 lb. Silicon
16 lb. Carbon
2 lb. Phosphorous
2 lb. Sulphur

In a Bessemer vessel, bring the iron to a boil and refine. Mix in alloying elements, achieve a molten state, and pour into ingots. Makes one ton. Serves 4,000.

As every cook knows, ingredients vary in quality and exact measure from batch to batch of the same recipe, especially when the recipe calls for many ingredients whose relatively small amounts come to be taken as more suggestive than precise. And when some stainless steel is made from virgin ore and some from left-over scrap iron, as it often is, the quality of the ingots can be even more uncertain.

Whether one batch of stainless steel is better or worse than another can depend more on the size of the pinches and dashes of the spices like sulphur that are used than on the quality of the main ingredients. When a large nuclear test reactor was under construction in the 1970s, engineers and metallurgists decided to call for fresh rather than recycled iron and some precise amounts of the spicing materials in a batch of stainless steel from which certain critical replacement reactor-core components would be manufactured. By making a special batch of steel they hoped to improve the mechanical properties of the parts in much the same way that the early American Liberty Bell founders altered the recipe for its metal in order to achieve a better tone. Unfortunately, the new reactor steel was not more but less tough than that from the original mix. Thus are the frustrations of working with complex alloys, but at least engineers and metallurgists gained another lesson in what *not* to do.

The manufacturer of our knives most probably did not use homemade stainless steel, but rather like the makers of common nuclear power plants, sent out for it and had it delivered. Whether the steel in our nine knives all came from the same batch is not known to me, but if some of it came from a cook's mistake, that could explain the cracks in some blades but not in others. Furthermore, if the cracked blades were indeed from a bad batch of ingots, that might give me less reason to suspect that the uncracked knives were at all unsound, for I might reasonably assume they were from a good batch, as I would expect a normal set of flatware to be.

The manufacturing process of the knives would be critical, and their individuality suggests that they experienced somewhat different processes. Perhaps they started as short blanks cut from long rods of steel. The blanks might then have been shaped into knives by pounding one end into a handle and by rolling or hammering the other end flatter into a blade. Such hammering or rolling, perhaps with one of several high-speed mechanical hammers under which an operator or a conveyor belt manipulated the blanks into blades, might possibly have been done with the steel very hot to the touch but far from melting, a process known metallurgically as "cold working."

After such forging the knives were probably heat-treated, a process of heating to a very high temperature and then cooling at a specified rate, perhaps in oil, to achieve a desired temper. The knife might then have been heated again to a metallurgically low temperature in order to remove any brittleness in the steel, and once again cooled. Among the finishing touches would have been sharpening and trimming the blade to its final shape, cutting a serrated edge into the side, stamping the blade with

STAINLESS
JAPAN

or a variation thereof, and polishing.

Too much cold working hardens steel, and harder steel cracks more easily than soft. Perhaps some of our knives, the cracked ones, were overworked by an overzealous hammer operator, or perhaps they were stamped too hard by a super-patriot. Another source of cracks could have been a slip of the polishing wheel or serration cutter or any such tool capable of introducing a nick or nucleus from which a fatigue or other type of crack could get a head start and grow prematurely. Or the knives may have been cooled unevenly or too rapidly, and this could have left residual manufacturing stresses in the blade.

This last is a very likely explanation because cracks introduced in such a way would more likely affect a group of knives rather similarly and would cause cracks to grow to a certain size quickly and then to become quiescent. Since I have noticed no appreciable change in size of the blade cracks in the dozen or so years since their discovery, any growth of the cracks due to mechanical fatigue processes would be too slow to account for the total degree of damage observed. Although the cracks might continue to be sharpened and extended by repeated use of the knives, the growth should be imperceptible for the foreseeable future.

There are still other possible explanations for the origin of the cracks in our knives. Sometimes stainless steel develops cracks due to the incompletely understood mechanical-chemical process known as stress-corrosion cracking. Three conditions are necessary for such cracking to occur: a critical intensity of stress, a corrosive environment, and a degree of sensitization during manufacture that renders the object susceptible to cracking. The knives certainly could have been stressed during cooling or cutting at the crack locations, and their proximity to common table salt could provide the deleterious chemical environment of chloride ions that can be so corrosive to stainless steel. (Stainless steel is actually a misnomer, for the metal corrodes readily in salt water.)

Some of our knives could have been sensitized by unusual heating and cooling during manufacture, while others might not have been. The prolonged exposure of the knives to salty leftovers after an all-night party might have precipitated cracks in the sensitized knives only. Subsequent rinsing and use could have worked the undesirable saline solution out of the interstices of the opening cracks and arrested the cracking. Or the cracks may have penetrated into an unsensitized region of the metal and arrested themselves as far as stress corrosion was concerned. Under my wife's photographer's loupe it can be seen that the cracks are clearly jagged and follow a zigzag path across the knife blade. This is characteristic of stress-corrosion cracks, which progress by wandering between the adjacent angular grains of metal rather than following a mathematically straight path through them.

All in all the evidence seems to indicate that the cracks in our knives originated in a self-arresting process of stress-corrosion or heat-treatment cracking and have grown imperceptibly, if at all, through fatigue. If this is so, new cracks are not likely to develop in the uncracked, and presumably unsensitized, blades, and they can be used unconditionally with confidence. The cracked knives that we have been using for all these years can also be kept in service, but they should be used with less vigor and watched for any accelerated crack growth signaling a new stage of cracking.

Watching cracks is not the only way to monitor their possible growth. Two knives, one with and one without a crack, when tapped against a glass of water or a cutting board will have a different ring. A crack will make a knife less stiff and cause it to vibrate at a different frequency from an uncracked knife. The human ear can easily distinguish between a cracked and an uncracked knife in our set and could detect significant changes in the lengths of the existing cracks. Like baseball players who tap their bats on home plate at every at-bat, we could but we refrain from tapping our knives on the dinner plate at every meal.

Unlike our knives, which have not developed new cracking problems for some years now, the Grumman Flxible buses continued to develop new problems of cracking in different areas of the structure. And that is why the buses were ultimately taken completely out of service in New York City. And after a mechanical analysis not unlike the one conducted on the knives was finished, the legal analysis began. And that promises to take a lot longer to resolve, for loopholes in contracts and chinks in legal armor can be harder to declare benign than cracks in steel, and the mettle of lawyers does not appear to be prone to exhaustion by fatigue.

While the concern for our cracked knives may have been an imaginative domestic exercise, the analysis involved is not much different than that which might follow from the discovery of cracks in a major engineering structure. However, unlike the speculation about the origins of the metal and the processes of manufacture of the knives, the investigation of cracks in a bus or, especially, a nuclear power plant should generally find records to answer more definitively the kinds of questions raised by the knives. With such records more definitive analyses of cracks and their potential dangers can be made, and the reports of these analyses can provide lessons and caveats for the future.

12

INTERLUDE:
THE SUCCESS STORY
OF THE CRYSTAL PALACE

Innovation in engineering, as in everything, involves risk and is an invitation for something to go wrong. But it does not follow that innovation *must* lead to failure. And because there have always been dramatic engineering projects that have proved the nay-sayers wrong about this or that daring new design, today's engineers are not acting irresponsibly when they want to use an untried material or structural design to build a bridge longer or a skyscraper taller than any one extant. They are merely following in the tradition of the great nineteenth-century builders of daring structures that outlived their opponents. And one of the most ambitious and innovative structures of the Victorian era was not a bridge or a tower but the vast building constructed to house the Great Exhibition in London in 1851. The story of the Crystal Palace is a fascinating one that bears repeating, for it shows that no matter how innovative an engineering structure may be and no matter how many opponents it may have, the proof is in the putting up and in the testing of it.

Joseph Paxton was born in 1801, the son of a farmer in Bedfordshire. As a young man he became a gardener, employed by the Duke of Devonshire, and by 1826 Paxton was superintendent of the gardens at Chatsworth, the Duke's Derbyshire estate. Paxton also displayed a special talent for structural design, and by 1840 he had built a greenhouse enclosing an acre of ground. This Great

Conservatory at Chatsworth was considered a contemporary marvel, and Paxton was firmly established as an engineer in practice and in spirit if not in name, for his Great Conservatory was to serve as a model for the Great Palm House in Kew Gardens, the Royal Botanic Gardens near London, and he was to go on to create many other structures of his own.

It was into one of the heated tanks at Chatsworth that Paxton first placed a cutting of a giant water lily obtained from Kew Gardens. The seeds had been brought back from tropical British Guiana in 1837, but the plant did not thrive at Kew. Under Paxton's care, however, the plant developed the huge leaves and beautiful flowers characteristic of such lilies. He named the flower *Victoria regia* (now properly called *Victoria amazonica,* since the plant had been so named before Paxton's involvement with it) and presented a bud to the queen.

As the lily continued to grow, Paxton designed a building especially for the plant, patterning the structure of the lily house after the structure of the lily leaves themselves. He once demonstrated the strength of the natural ribbed structure by placing his young daughter on one of the floating leaves, which measured five feet in diameter. It easily supported her weight without sinking, and it later became a Victorian fad to be photographed on one of the giant water lily pads with the characteristic lip around the edge that makes them look like baking dishes for enormous quiches. Paxton observed that the large leaf owed its strength and stiffness to the geometric pattern of ribs and crossribs on its underside, and he took that as a model. The result was a lily house at Chatsworth measuring 47 feet by 60 feet (about 3,000 square feet), with a glass roof resting on wooden beams set across iron girders supported by iron columns. This light and airy building eventually provided the idea for the Crystal Palace, which would measure over 400 feet by 1,800 feet (almost 750,000 square feet) and shelter the 100,000 exhibits of the Great Exhibition of the Works of Industry of All Nations, as the first world's fair was officially designated.

Just as the world of commerce was ready at mid-century for the first international exhibition of the works of industry, so was the world of technology ready to build the Crystal Palace. The British government had repealed the century-old glass excise tax in 1845, removing any fiscal impediment to using almost 300,000 panes of glass in the building. The United Kingdom was producing about five million tons of cast iron and wrought iron annually, more than 1,000 times the amount required for the exhibition building, which at 4,500 tons was still an enormous quantity. And although the scale of the Crystal Palace was indeed grand, the engineering experience gained in developing Britain's railroad system, which included hundreds of iron bridges, provided the knowledge about the strength of materials necessary to execute the bold design. (Even though at the time bridges were failing at an alarming enough rate for a Royal Commission to have been appointed to look into the use of iron as an engineering material, there appears to have been sufficient confidence in using iron in a gigantic static structure that would not be subjected to the pounding of railroad wheels and the higher and higher loads of evolving rolling stock.) Nevertheless, the Great Exhibition came to be housed in the Crystal Palace only at the eleventh hour.

An exhibition of international scope, featuring the application of art to industry, was first suggested by Henry Cole, a public servant and art patron whose indefatigable energies and interests made him a driving force behind Victorian architecture and industrial design. Unlike his contemporary John Ruskin, who did not look favorably upon the common products of the Industrial Revolution, Cole's vision was to combine the fine arts and engineering. Thus Cole's idea of a Great Exhibition was a natural extension of the views he expressed before the Society of Arts in 1847:

Of high art in this country there is abundance, of mechanical industry and invention an unparalled profusion. The thing still remaining to be done is to effect the combination of the two, to wed high art with mechanical skill.

Prince Albert immediately embraced the idea of a Great Exhibition and agreed that Hyde Park would be the best site. Early in 1850 the prince became chairman of a Royal Commission to promote the project, and soon a building committee was appointed. This committee envisioned a temporary structure covering sixteen acres and announced an open competition to select the design. However, the committee found none of the 245 entries acceptable and proceeded to cannibalize them for what it considered the best features to produce its own design. A rendering of the committee's "camel" of a building was published in the *Illustrated London News* in June and was panned immediately by the *Times*.

Critics described the proposed structure as a "vast pile of masonry" that they feared would never be removed and would become a "permanent mutilation of Hyde Park." Indeed, the committee's malproportioned behemoth would have required the laying of fifteen million bricks and the construction of a dome two hundred feet in diameter, considerably larger than that of St. Paul's Cathedral. The mortar would not have been expected to be dry in time for the opening of the exhibition, then less than a year away.

Meanwhile, the exhibition itself and its location were debated in Parliament. Xenophobic opponents feared the effects of foreign competition on the sale of British goods at home and abroad, while others cited less commercial concerns, including overcrowding, rowdiness, and disease. One particularly vocal opponent was Colonel Charles Sibthorp, a protectionist who also had opposed the Public Libraries Act of 1850 because, among other reasons, he "had not liked reading at all." His most celebrated objection to the exhibition site involved a small clump of elm trees marked to be cut down. Reason prevailed in Parliament, however, and the protests of Colonel Sibthorp and his sympathizers were for the moment suppressed, though the environmental impact issue never did entirely disappear.

Paxton did not enter the original competition because, according to his own account, he took it for granted that the building committee would select a suitable design without his help. But he was disappointed in the proffered plans, which were being discussed more or less publicly, and wondered if it was too late to put forth an idea. Although he approached exhibition officials only a fortnight before the building committee's final choice was formally to be announced, Paxton persuaded them to allow another entry. The commitment was obtained on June 11, 1850, and on that day Paxton left London for the Menai Straits in northwestern Wales to see the third iron tube girder of the Britannia Bridge set in place. He was distracted during a board meeting that he attended while away from London by the thought of a building for the Great Exhibition, and it was at that meeting that he made on a blotter his famous sketches of what was to become the Crystal Palace. During the following week, with the help of Peter Barlow, a railroad engineer, the columns and girders were sized and the design completed. When Paxton showed them to Robert Stephenson, whom he met on the way back to London, the bridge engineer was enthusiastic.

At first the building committee pooh-poohed the new proposal, but it gradually gained support. And after a rendering of the design was prematurely published in the *Illustrated London News* of July 6, the committee soon abandoned its own design and unanimously embraced Paxton's "ferro-vitreous"—iron and glass—concept. Among the advantages were the building's extreme simplicity, the speed with which it could be assembled, its absence of internal walls, and the fact that the materials could be reused. The economic advantages of lower construction costs and high salvage value seemed to clinch the design's choice. (Indeed, the Great Exhibition—unlike virtually all subsequent world's fairs and international exhibitions—made a handsome profit, certainly in part because of the economy and smoothness of construction of the Crystal Palace.)

However, the fate of the elms in Hyde Park still cast a threatening shadow, and Paxton added to his original design a central transept that would enclose the ninety-foot-tall trees. Construction began even without a firm agreement with the contractor, Messrs. Fox, Henderson, and Co., and a contract was not signed until more than a month later. The price agreed upon was £79,800 to erect and later dismantle the building, with the contractor owning the materials. The final cost of the Crystal Palace, including modifications such as the transept and other additions, was £200,000. Still, this works out to about £25 per hundred square feet of covered ground, a bargain even in the mid-nineteenth century.

Workers fenced in the construction site in August, with the same wooden planks that would be used later for the floors and galleries. Completion of the project was scheduled for January 1851, which allowed just over twenty weeks to enclose with massive amounts of iron and glass an exhibition floor whose requirements had grown to nineteen acres. The ground was leveled, foundations and iron drainpipes were laid, and the first column was erected on September 26. Construction proceeded quickly, by the light of bonfires at night, and on one Saturday Paxton reported seeing two columns and three girders put up by two men in only sixteen minutes. What was to be the first large and truly significant building constructed of metal and glass would be erected in only seventeen weeks!

The mathematical regularity of the proportions of the Crystal Palace helped simplify construction. Paxton determined the basic unit of length for the building not by some aesthetically based "golden section," but by the requirements of the exhibition space and a fundamental technological constraint: in 1850, sheets of glass longer than about four feet were not only very costly to manufacture but also were very unwieldy to install. Thus, panes forty-nine inches long were mounted on the roof in a gently sloping "ridge-and-valley" pattern that was not only pleasing to the eye but also provided good drainage. To simplify construction, workers installed the glass as they moved along in trolleys that rolled on wheels set in the grooves of "Paxton gutters." These were installed in the valleys to carry rain and condensation to the hollow iron columns that doubled as drainpipes. The gutters were placed at eight-foot intervals, a length dictated by the combined span of two sheets of properly inclined glass. Roughly three times this dimension, or about twenty-four feet, was a convenient length for the cast-iron girders, and their supporting iron columns were placed at twenty-four-foot intervals. Thus, twenty-four feet became the basic unit of scale in the plan for the entire Crystal Palace.

Wide interior "avenues" that stretched the length of the building were spanned by wrought-iron trusses forty-eight-feet long. And the spectacular Central Avenue was spanned by trusses three-units, or seventy-two-feet long. The arched central transept was also seventy-two-feet wide. This bare and simple geometric regularity no doubt contributed to the structure's graceful appearance, just as the repetition of basic units of glass or marble facing work so effectively in modern architecture. The seventy-seven twenty-four-foot units along the south facade of the Crystal Palace were also numerous enough to distract all but the most observant and counting eye from the fact that the central transept was not central at all, but was off center by one unit to accommodate Colonel Sibthorp's elms.

The uncommon depth of the girders and trusses—roughly three feet—added to the grace of the Crystal Palace when a visitor looked down one of the long avenues. And in Paxton's usual custom, this aesthetic feature served a structural purpose as well: attaching the trusses and girders to the supporting iron columns not only at the top but also at their relatively deep bottom provided a great stiffness against wind and other lateral forces. The structural advantage of the deep girders can be appreciated by comparing the flimsiness of a card table having thin folding legs with the rigidity of an old kitchen table whose top is supported on deep wooden side beams attached solidly along their entire depth to firm legs.

The construction techniques, like the building design, also speeded up progress. For example, workers used special machines developed by Paxton to cut several wooden sash bars simultaneously out of a single plank, grooving and beveling their edges at the same time. Workers used circular saws to notch and bevel the ends of the sash bars, which would hold the panes of glass in place, and pierced nail holes with revolving augers driven by a steam engine. Altogether 600,000 cubic feet of timber were used in the Crystal Palace, including twenty-four miles of Paxton gutters.

Increasing the scale of something from a relatively modest lily house to a grand exhibition building can be tricky business, for in engineering as in nature bigger is not necessarily a better or even a good idea. Thus, besides the environmentalists who feared the ravishment of Hyde Park, the health officials who worried about sanitation and disease, and others who worried about fire, comfort, and crime among the millions of visitors expected at the Great Exhibition, there were those who simply did not think the structure itself was safe. When the King of Prussia inquired about the safety of visiting London during the Great Exhibition, Prince Albert wrote an ironic letter of caution. He recited all the dire prophecy and confessed that "Mathematicians have calculated that the Crystal Palace will blow down in the first strong gale; Engineers that the galleries would crash in and destroy the visitors."

While questions about the strength and stability of the Crystal Palace arose from the beginning, its design and construction made no compromises with safety. Wherever the girders and trusses did not provide sufficient lateral stiffness, slender diagonal rods were installed in cross patterns. These rods were used extensively in the arched central transept and their oblique pattern broke up what might even have been the monotony of the great space. Workers tested *all* cast-iron girders as they arrived at the construction site using a hydraulic press made for the purpose and achieved a degree of quality control rarely matched even today. They also tested the double- and triple-length wrought-iron trusses, but needed to test only one of each type since the wrought iron by its nature could be counted on to be of uniform quality as surely as the cast iron could not.

Criticism continued throughout construction, however, and nay-sayers warned that wind and hail would bring the glass box down, or that heat and humidity would make it unbearable in the London summer. Neither happened, though one parody in *Punch,* which was generally supportive of the great building that it christened the Crystal Palace, was a biting piece narrated by a cucumber who knew what it was like in a hot glass house on a steaming July day. But the Crystal Palace withstood the elements and proved to be as cool (thanks to canvas suspended over the roof and adjustable louvers in the walls) and dry (thanks to the Paxton gutters) as one could wish. But, said the predictors of doom, if the forces of nature would not be the undoing of the Crystal Palace, the masses of visitors would.

Extensive galleries were planned to provide an additional 200,000 square feet of promenade and exhibit space, but the elevated walkways were attacked as unsafe months before the opening of the Great Exhibition. After all, during this time iron railway bridges were failing at a rate of almost one in four, and suspension bridges were collapsing under marching soldiers. The safety of the Crystal Palace galleries had yet to be demonstrated. As a test, a twenty-four-foot-square section of gallery was constructed just off the floor on four cast-iron girders. Inclined gangways of wooden planks served as approaches to the test floor. Queen Victoria, on one of her inspection tours, witnessed with her entourage the following proof test, as reported in the *Illustrated London News* of March 1, 1851:

> The first experiment was that of placing a dead load of about 42,000 lb., consisting of 300 of the workmen of the contractors, on the floor and the adjoining approaches.
>
> The second test was that of crowding the men together in the smallest possible space; but in neither case was there any appreciable effect produced in the shape of deflexion. So much for dead weight.
>
> The third experiment—which was that of a moving load of 42,000 lb. in different conditions—consisted in the same party of workmen walking first in regular step, then in irregular step, and afterwards running over the floor, the result of which was equally satisfactory.
>
> The fourth experiment—and that which may be considered the most severe test which could possibly be applied, considering the use to be made of the gallery floors when the Exhibition is opened to the public—was that of packing closely the same load of men, and causing them to jump up and down together for some time: the greatest amount of deflexion was found to be not more than a quarter of an inch at any interval.
>
> The third experiment was then repeated, substituting, however, the Sappers and Miners engaged at the works, for the workmen of Messrs. Fox, Henderson, and Co.; and this last trial, which was quite as satisfactory as the others to all present, is represented in our illustration.

The specimen gallery withstood it all, and the *Illustrated News* went on to hope that the fears about the safety of the building had "by this time, been entirely rejected from the minds of those who have gone so far as to predict that the 1st of May would but prove fatal to the thousands who will enter the great Industrial Palace on that occasion."

The successful testing of the galleries served also as the basis for a Tenniel cartoon in *Punch,* which declared itself a proud "Sponsor of the Crystal Palace." In the cartoon the specimen section of the gallery is shown supporting a great ball labeled "THE WORLD" and Punch himself is beaming atop it. The gallery floor shows not the slightest strain and the crowd witnessing the event is all smiles, with their hands raised in triumph and their hats off to the Crystal Palace.

The Crystal Palace was unique not only in its structural details but also in its maintenance and decoration. The floor, for example, was laid with a space between the boards so that dirt and debris could fall or be swept into the cracks, preserving a neat and dustless promenade. Floor-sweeping machines were originally to be used to push the dust into the half-inch spaces between the boards, but they proved unnecessary as women's dresses accomplished the same end. Small boys were employed to crawl beneath the floor boards and collect bits of paper that might otherwise accumulate and present a fire hazard.

All the decorative and ornamental details for the building were under the direction of Owen Jones, known as "Alhambra" Jones because of his great knowledge of Moorish architecture. In painting the structural elements, Jones applied his "science of color." His contemporaries did not universally applaud the result, but it is difficult to judge today because original hand-colored prints have long since faded. However, the colors have often been described in words.

Pale blue was the predominant color of the interior vertical metalwork, and this is believed to have enhanced the feeling of open space. The underside of every girder was painted strong red, a color repeated in screens against which many of the exhibits were to stand. Yellow was used on molded details and highlighted the fluted portions of the otherwise blue columns. All in all it must have been a striking palette. A further touch of color was added on the exterior by displaying flags of all nations on a thousand poles around the periphery of the roof. (Sir Charles Barry, one of the organizers of the Great Exhibition, suggested this.) The flags had the remarkably pleasant effect of breaking up the monotonous straight line of the long roof in a most appropriate way.

Another example of Jones' attention to detail was an electric clock with a face twenty-four feet in diameter located above the south entrance of the central transept. The clock might have dominated and disfigured the transept had Jones not abandoned the traditional circular arrangement of the hours for a face that exploited the design of the transept itself. The numerals were arranged in a semicircular pattern on the transept's radiating structural components. Instead of a single hour hand that swept in a circle every twelve hours, the clock had an hour hand (really two hour hands) that looked like a propeller. It revolved only once every twenty-four hours, with one blade of the propeller indicating the hour at any given time. The minute hand was similarly designed.

If anything did not go off like clockwork in the Crystal Palace, it is hardly remembered. Just as we seem to remember nothing but the failure of an ill-fated design, forgetting what might have been its successful innovations, so we seem to remember nothing but good of those designs that succeed. "He built wiser than he knew," wrote journalist Horace Greeley of Joseph Paxton, and the Crystal Palace seemed to succeed even beyond everyone's hope and expectation. The building itself stole the show from the exhibition's myriad displays of manufactured goods.

On May 1, 1851, Queen Victoria opened the Great Exhibition with much pomp and circumstance before an assembly that included numerous foreign officials and functionaries. More than six million people were to visit the exhibition during the 141 days it was open. (It was closed on Sundays.) The busiest day saw more than 100,000 visitors, with 90,000 people in the building at a single time. The galleries apparently never trembled, and there were no panics about the safety of the structure. The Queen herself returned to the Crystal Palace about fifty times before the Great Exhibition closed on October 15, 1851, and she seemed never to tire of spending hours methodically touring the exhibits. A retrospective entry from the Queen's journal for the closing day reads: "To think that this great and bright time is past, like a dream, after all its successes and triumphs."

Although the Crystal Palace was supposed to have been dismantled after the exhibition so that Hyde Park could be restored to its unimproved state, officials seriously considered leaving the scaled-up lily house where it was. Some wanted to turn the structure into a "winter garden" where people could ride and walk among flowers during the dreary days of the long London winters. The costs of adapting the building for permanent use in Hyde Park, as opposed to dismantling it and reerecting it elsewhere, were compared in great detail. But Colonel Sibthorp, perhaps remembering the elms outside the transept that were cut down, opposed permanent installation. Various proposals for relocating the building were forthcoming.

Among the daring proposals for reusing the columns and girders was Prospect Tower, conceived to be a thousand feet tall. This tower would certainly have been an economical use of ground, as its designer pointed out. The tower would have sported a clock forty-five feet in diameter with numerals ten feet high, and proponents were sure its glass exterior would withstand the great forces of the wind. This vision of the modern skyscraper was aesthetically if not structurally a century ahead of its time and would compare favorably with the designs of today. However, it would certainly have taxed the elevator technology of the early 1850s, and it would surely have been a bolder leap in structural engineering than even the Crystal Palace and not likely to have succeeded as well.

Yet, although the true skyscraper did not really come into its own until the twentieth century, the Crystal Palace prefigured it in many important ways. The light, modular construction ingeniously stiffened against the wind is the essence of modern tall buildings. And the innovative means by which the walls of the Crystal Palace hung like curtains from discrete fastenings, rather than functioning as integral load-bearing parts of the structure, is the principle behind the so-called curtain wall of many modern facades.

The Crystal Palace inspired much contemporary architecture, as the idea of international exhibitions spread quickly throughout the world. In 1853 New York hosted a world's fair in a cruciform iron-and-glass "palace" topped by a dome 168 feet high. Here Elisha Otis demonstrated a new safety device for elevators. He ascended in an elevator cage to dangerous heights above the floor and, before a gasping audience, cut the supporting rope. The elevator started to fall but was stopped by Otis' simple gravity-activated locking device that gripped the guide ropes. This was a milestone in mechanical engineering that, like the civil-engineering milestone of the Crystal Palace itself, was essential to the development of the true skyscraper.

Although the Crystal Palace was doomed in Hyde Park, Paxton had other plans for the building that had earned him knighthood. Dismantling began in the summer of 1852, and the columns and girders, gutters and glass, were transported to two hundred wooded acres atop Sydenham Hill, south of London. Paxton raised over half a million pounds to purchase the Sydenham site and the building's construction materials.

The Sydenham Crystal Palace was to be more than a re-creation of the original structure, however. The roof was vaulted along its entire length, and the central transept was greatly enlarged and doubled in width so as not to be overwhelmed by the new roof. The enlarged transept, in turn, demanded the addition of two stories and two end-transepts for balance. The final cost of the project, which included extensive gardens and fountains, was £1.3 million.

Part of the extra expense was for two tall water towers built to supply the elaborate fountains that were intended to rival those at Versailles. Not only did the towers store water, but also the South Tower housed the Crystal Palace Engineering School attended by, among others at the turn of the century, a young Geoffrey de Havilland. Years later Sir Geoffrey, the airplane manufacturer, recalled fondly in his autobiography that the building at Sydenham provided myriad diversions not only for engineering students but also for all Londoners.

Although fire destroyed the north transept in 1866, the remaining asymmetrical memorial to Joseph Paxton and the original Crystal Palace withstood wind and hail, if not waning interest, for many more years until another fire destroyed the entire structure in 1936. The two water towers remained standing until they were demolished in 1940, presumably because of concern that they might serve as beacons for enemy bombers looking for London. Today a television transmission tower stands on the site, and in its shadow is a bust of Sir Joseph Paxton atop a cracking stone column. Throughout nearby London, and throughout the world, architectural descendants of the Crystal Palace abound. But no matter how tall, they do not seem to approach the greatness of their progenitor.

While the Crystal Palace holds a secure though not unassailable place in the history of both engineering and architecture, whether it represented a triumph for either discipline was hotly debated in 1850. Although Paxton had been designing and overseeing the construction of parks, gardens, greenhouses, and conventional buildings for over two decades, his lack of professional training in either discipline caused his plans to receive a cool reception in some quarters. People, like the buildings they make, can have their failings. Distinguished members of the Institution of Civil Engineers were among the leading prophesiers of the collapse of the original Crystal Palace, and Paxton was never to receive the Royal Gold Medal in Architecture.

Victorian architects generally found the Crystal Palace lacking in a strong sense of form or organic integrity. They argued that Paxton's use of repeated modules was arbitrary and lacked artistic motivation or restraint. Indeed, there may be some validity to the criticism, for Paxton's first design was based on twenty-foot modules. He increased the size to twenty-four feet only when he learned that this was the minimum width for an exhibitor's stall. And the overall length of the building was coincidentally and whimsically 1,851 feet—the year of the exhibition—as Charles Dickens observed in his popular weekly, *Household Words*. But regardless of what professionals thought at the time, Paxton's Crystal Palace captured the hearts of London and the world in the mid-nineteenth century as few buildings or structures seem to have since.

Because Paxton was not steeped in the traditions of either engineering or architecture, he approached design problems without any academically ingrained propensity for a particular structural or aesthetic style. He solved the problems of housing a giant water lily and a great exhibition alike with buildings that departed both from conventional methods of construction and architectural traditions. In short, Paxton, in his professional naïveté, struck out in brilliant new directions that produced models for the architects and engineers of the next century. The Crystal Palace was the first large and truly significant building to be made of metal and glass, the first major building to use outer walls that provided no structural strength, and the first building constructed using prefabricated, standardized units that were shipped to the construction site for rapid assembly. These practices are now commonplace in many large construction projects. Among the architectural breakthroughs of the Crystal Palace was the building's use of colossal space. The repetition of structural units—enhanced by interior decorator Owen Jones' color scheme of "yellow rounds, blue hollows, and red flats"—was a clear forerunner of much of modern architecture. The building also stands, or stood, as a success story in construction management. Although built in what may seem to have been a simpler age, the Crystal Palace was not without many of the same complications that can delay modern construction projects and cause them to fall years behind schedule. The project involved major planning, financial, management, and labor components. The amount of materials that had to be ordered, manufactured, delivered, processed, and erected was enormous even by today's standards. Social and political obstacles existed as well, for although an environmental impact statement was not required in 1850, Paxton still had to accommodate many objections, including those that today would be raised by environmentalists.

As the roles of architects and engineers moved further and further apart in the latter half of the century, the Crystal Palace served as a symbol of what could be but would not be again until the mid-twentieth century. Of course, conservatories and greenhouses continued to be built in the Paxton tradition. But for buildings designed to house solid institutions cultivating and preserving money, art, knowledge, and other relatively immortal commodities, heavy brick, stone, and cast-iron facades were the preferred style.

Perhaps the centennial of the Crystal Palace contributed to its architectural reincarnation. In 1951 seemingly countless exhibits were staged and books published on the Crystal Palace, and Joseph Paxton and his building began to be favorably reappraised. In the 1950s Lever House and the Seagram Building, both characterized by the non-structural "curtain wall," rose in New York City to epitomize the concept of the tall glass box that has since become ubiquitous. Indeed, Mies van der Rohe, who like Paxton had no formal architectural training, had actually anticipated in 1921 the multitude of successors to his Seagram Building with his then unrealized "Glass Skyscraper," whose facade was reflected in its own curved forms. And his apartment buildings in Chicago, with their strong exterior structural details carried to an extreme and using superfluous steel beams and columns for purely decorative effects, helped to reawaken the sense of the technological roots of architecture implicit in the Crystal Palace. The ultimate example of this concept may be the exposed structural and mechanical elements of the glass-walled Centre d'Art et de Culture Georges-Pompidou in the Beaubourg section of Paris.

The influence of the Crystal Palace is especially strong today, as architects include large atriums and open public spaces in their designs of corporate headquarters and other urban buildings. The IBM Building in New York City has a four-story greenhouse, complete with bamboo trees green in winter beneath its saw-toothed roof, which cannot help but evoke Paxton's patented ridge-and-valley design. And the legacy of the Crystal Palace lives on today in more than an atrium in at least two major construction projects in this country. The new Infomart in Dallas is the world's first large-scale center for marketing computer products, and its appearance is as faithful to the original Crystal Palace as architect Martin Growald could make it. He even tried to match the Victorian color scheme of the building that housed the Great Exhibition, but unfaded prints and intact artifacts of the nineteenth-century structure are hard to come by.

Growald's Crystal Palace sits on the Stemmons Freeway among the more boxy and opaque buildings in the wholesale market complex known as the Dallas Market Center, built by the Trammel Crow Company. The original concept for a computer market building called for a conventional structure with a rectangular shape, a granite and glass curtain wall, and a rooftop glass vault. When Growald was asked to design the building, he suggested to Trammel Crow that the computer market building be diametrically opposed to the other ones in the complex. The architect sees his design for the Infomart as ideal to serve as an exhibition and marketplace for computers, which he sees as the instruments for a continuation of the Industrial Revolution. However, the construction of a modern Crystal Palace proved to be a lot more expensive than one of today's more familiar market buildings, and the project was delayed while comparative economics were argued and wrestled with. The $85 million first phase, consisting of six stories of a projected fifteen, was finally approved and ground broken in May 1983. The building was ready for occupancy in January 1985. While this is a longer time than it took to complete Paxton's Crystal Palace, which did not need so many electrical outlets for computers, the Infomart was a relatively trouble-free construction project by modern standards.

The Crystal Palace of 1851 has also inspired the New York Convention Center, but its construction has been as beset by as many problems and delays as Paxton's and Growald's buildings escaped, though the re-erected building at Sydenham, whose design went beyond the technology of the original Crystal Palace, may be said to have prefigured what has happened with the New York project. Groundbreaking for the convention center took place in 1980 and construction was to proceed on a "fast track" for a 1983 opening. The sixteen-acre structure is of the same scale as the 1851 building that was erected in Hyde Park, but its design is strictly twentieth-century. Instead of girders bolted to columns, the New York Crystal Palace will have the world's largest space frame. Two expansion joints will be provided in the roof of the six-block long structure, which engineers believe will behave more like a bridge than a building, and the lessons learned from wind blowing across bridges were used to design features to counteract the uplifting forces. The space frame will be put together in Tinker-toy fashion from steel rods that fit at various angles into steel hubs full of holes.

The project was priced at $375 million when the design was revealed in late 1979, but in early 1983 the news broke that defects were discovered in many of the hub castings that were to be located at the eighteen thousand nodes of the space frame. X-ray tests of completed nodes turned up numerous cases of voids and hairline cracks in the metal that would make the components unable to carry as much load as they were expected to in the original design. Early reports projected that this unexpected trouble would add $5 million to the cost of the building, and one plan to keep the construction schedule from slipping too much called for using larger, defective hubs to substitute for some of the smaller ones. In this way the twelve thousand parts that had already been cast, some sixty percent of which were weaker than specified, would not have to be discarded completely. This remedy was eventually rejected, however, and a totally new order was placed to have all hubs recast—at a cost of over a million dollars and a delay of more than a year. In the meantime, the tall, massive, welded-steel columns were the dominant features on a quiet construction site that looked like an abandoned garden whose pollarded trees were planted too far apart in an arid field.

As soon as a completion date had been set for the Convention Center, space beneath what would be Manhattan's gigantic skylight began to be committed. Before construction problems became front-page news, more than fifty conventions were booked in the blocks-long dream-structure for the last half of 1984 alone. The consequences of a delay have been disastrous not only for the lost revenue and lawsuits from conventions without a home, but also for the lost new bookings. After all, who is to say for sure when the troubled convention center will finally open? The governor of New York ordered a report on the delayed project, and when it came to him almost three months late it put the completion date in the summer of 1986. The projected cost was increased to $500 million, and that did not include the hundreds of millions of dollars of convention business lost to New York City for the years 1984 and 1985 that the center was to be opened and operating.

Fast-tracking was blamed for many of the delays experienced in the construction of the New York Convention Center. In this method of construction, where certain components are put in place before the whole design is complete, options can be limited rather than opened up, according to opponents of the method, and fast-tracking makes it impossible to know the final cost of a structure at the outset. Proponents counter with the claim that, by overlapping the design and construction steps of building, the system provides some protection against inflation, and can get a project finished sooner than if every design detail is completed before any construction begins. The fast-track method has been given good marks in Washington, D.C., where it was used in the construction of a convention center, and in San Francisco, where it was used in building the Moscone Center. The problems experienced in New York may not have been due to fast-tracking per se but to problems unique to the management of a large public construction project and to the lack of quality control as demonstrated in the problem of the cracked and defective hubs.

The difficulty of producing reliable structural components by casting was known to the builders of the original Crystal Palace, and that is why they had *every* cast girder tested before they accepted it for delivery at the construction site. They also knew that wrought iron is more dependable, and thus they only tested a few of the girders of that kind, taking the test results to be representative. The cast hubs for the New York Convention Center were sampled and tested, with statistics providing the confidence that the inspection of some but not all individual parts was sufficient. But while sampling methods may have improved since the nineteenth century, the same inherent problems with castings remain, as the experience in New York has demonstrated.

Setbacks may seem to be the norm in construction, but as the historical example of the original Crystal Palace illustrates (the erection in fourteen months of the Empire State Building could serve as another positive case study), they are not inevitable. And even when problems do occur, they are often forgotten when the structure is complete. Innovation, whether it is in structural engineering design or in construction methods, can not only be a threat to success but also can cause the designers and builders to be more cautious in anticipating and eliminating trouble. Knowing they are bridging record distances or enclosing record volumes can call forth in engineers the attention to detail that made the Crystal Palace the symbol of success that it has come to be.

13

THE UPS AND DOWNS
OF BRIDGES

The first book of riddles published in English appeared in 1511, and it posed, among others, the question "What thing is it, the less it is the more it is dread?" And the answer was given as, "a bridge." Then, as now, bridge builders were apparently looking for ways to make their structures lighter for aesthetic and economic reasons, even as they were making them span ever-greater distances. And because of the psychological and structural struggles between less and more, the history of bridges is crisscrossed with interrelated examples of colossal failures and spectacular successes, providing many technological riddles, lessons, and surprises.

For over a hundred years now, the Eads Bridge has crossed the Mississippi at St. Louis in three chrome-steel arches while the Brooklyn Bridge has been suspended over the East River in New York from four steel-wire cables. These two bridges represent some of the earliest uses of pneumatic caisson technology and steel in bold new structural designs, and both stand as unequaled achievements of their era. They both involved several simultaneous leaps of engineering and hence invitations for something to go wrong. Yet, like the Crystal Palace, the Eiffel Tower, and other unique structures of the nineteenth century, these bridges are dramatic examples that great and daring engineering projects can indeed overcome whatever might go wrong and thus defy Murphy's Law.

Both construction projects encountered their share of setbacks, not the least of which involved the new materials they employed in critical structural components. James Eads actually adopted chrome steel, of which many metallurgists of the time were skeptical, only after he could not get satisfactory carbon steel parts for his bridge. And John Roebling's son Washington, who oversaw the construction of the Brooklyn Bridge after his father's untimely death, discovered—only after a certain amount of steel wire had already been spun into the bridge's cables—that inferior material was being provided by the supplier. Rather than undo what had already been done, Roebling increased the amount of wire that would make up the cables so that the weaker strands would be compensated for. That inferior wire remains in the cables today, a testament to the factor of safety concept, to the corrective measures that can be taken in design and construction, and to the fact that flaws per se do not necessarily lead to failure.

Those who doubted the structural success of the designs of Eads and Roebling were answered when the bridges were finished. The Eads Bridge was completed in May 1874, and progressively heavier railroad trains passed back and forth across it into mid-June. The final test seems to have occurred when an elephant crossed the bridge without hesitation, for it was popularly believed that such beasts had instincts that would keep them from setting foot on an unsound bridge. On the Fourth of July, 1874, a parade celebrated the completion of the bridge between southern Illinois and St. Louis, and in his speech Eads acknowledged and dismissed the doubts about the bridge's safety: "Yesterday friends expressed to me their pleasure at the thought that my mind was relieved after testing the bridge," he stated. "But I felt no relief, because I had felt no anxiety on the subject." But if Eads was confident his stone and steel structure would last as long as the pyramids, he was to grow as disappointed in its economic performance as the parade-goers were in the fireworks display that failed to produce the planned illusion of a "phantom train" crossing the bridge during the opening night festivities. Neither did the Eads Bridge earn the tolls that had been projected, and it never did recapture from Chicago the Midwestern trade routes that once passed through St. Louis and made it the Gateway to the West.

The Brooklyn Bridge, the "Eighth Wonder of the World," opened spectacularly in May 1883 to parades and fireworks. The pyrotechnic display was apparently as successful as the one in St. Louis was not, and the bridge came to be the economic success its backers hoped it would be. Unfortunately, the safety of the Brooklyn Bridge was not confirmed by the crossing of an elephant, and only a week after the grand opening a crowd of twenty thousand panicked and twelve people were trampled to death when a rumor of impending disaster spread across the bridge. The newspapers were blamed for creating public doubt about the safety of the structure. Yet today the the Brooklyn Bridge stands as a symbol of stability in an unstable world (even though in 1981 one of its diagonal cables snapped and cut down a pedestrian who was enjoying the structurally unnecessary but humanly brilliant scenic promenade elevated above both the bridge and the river traffic).

The history of suspension bridges, which illustrates so well the delicate balance that can exist between engineering success and failure, is curiously both more lore-filled and sporadically more forgotten than that of any other type of structure. An early suspension bridge over the Maine River at Angers, France, is said to have collapsed in 1850 under the marching step of five hundred soldiers, killing nearly half the troop. The soldiers' custom of breaking step when reaching a bridge is said to have stemmed from that accident, and the testing of the galleries of the Crystal Palace by trooping workers was prompted at least in part by the frequent failures of nineteenth-century suspension bridges under the massed movements of soldiers, cattle, and even crowds of people watching boat races. Such incidents may have been in the minds of those who advanced the theory that rhythmic dancing on the ill-fated skywalks of the Hyatt Regency Hotel caused them to undulate in ever-increasing amounts until they fell. While the hotel skywalks were much stiffer than a long suspension bridge or catwalk, dancing a bridge to destruction is not so farfetched an idea as at first it might seem. But hotel personnel and patrons supposedly found any bounce of the walkways no more unsettling than that of the thin, springy concrete galleries and balconies in other modern hotels and shopping malls.

Washington Roebling knew of the dangers of shaky walkways and erected a sign to that effect in a very prominent location at the entrance to the long, suspended catwalk to the towers of the Brooklyn Bridge when it was under construction. The sign clearly warned of the dangerous phenomenon of structural resonance, which could cause the walkway's motion to grow with each rhythmic footstep the way the arc of a playground swing grows with each well-timed push:

> SAFE FOR ONLY 25 MEN AT ONE
> TIME. DO NOT WALK CLOSE TOGETHER,
> NOR RUN, JUMP, OR TROT. BREAK
> STEP!
> W. A. Roebling, *Eng'r in Chief.*

These were explicit warnings against conditions that the structure's designer knew could lead to failure. As in many construction situations, rather than make the temporary structure overly strong, its use was restricted. There were no unfortunate incidents on the catwalks, and with its completion, the Brooklyn Bridge ushered in a new era of suspension bridges. And its permanent roadway was stiff enough not only to resist the twisting action that would later do in the Tacoma Narrows Bridge but also was set to a different drummer than the beat of soldiers' feet.

John Roebling lived in a time of suspension bridge failures, and his notable successes—the double-decked Niagara Bridge, the Cincinnati Bridge over the Ohio River, and the Brooklyn Bridge —owed the stability of their long spans under both traffic and wind loading to his understanding that the collapse of so many suspension bridges of his contemporaries was due to a deck that was not properly stiffened against those loads. But the successes of Roebling's designs stood more as symbols and encouragements than as lessons to future bridge engineers, who were presented with challenges to span wider rivers and bays with more economical structures.

Roebling himself rose to the challenge of spanning the Niagara River gorge with a radical design for its age, even though some of his contemporaries said it could not be done. Robert Stephenson, the great British engineer whose Britannia Bridge was built up of huge rectangular tubes of steel through which railroad trains could pass because he did not believe a suspension bridge could ever carry trains, wrote to Roebling about his proposed suspension bridge to carry railroad trains across the Niagara: "If your bridge succeeds, then mine have been magnificent blunders." The Niagara Bridge did succeed, but then so did Stephenson's "blunder." His design carried trains across the Menai Straits for over one hundred years. It was replaced by an arch bridge only after being destroyed by a fire in 1970.

Robert Stephenson could view his tubular bridge as a potential design "failure" (even though it was a structural success) because his contention in designing it had been that no structurally more efficient bridge could be built for the heavy, moving loads of railroad trains. His contemporary William Fairbairn actually blamed the success of Stephenson's tubular iron bridge, which did not even need the support of cables from the towers provided for them, for a number of "weak bridges" built after 1850, perhaps overconfidently, in the wake of the Britannia's success. Roebling's achievement in the Niagara Bridge provided at the same time a counterexample to Stephenson's claim about suspension bridges and a model for future bridge designers to attempt to surpass. Roebling used many of his own innovations from the Niagara Bridge—among them a deep, stiffened roadway and diagonal cables to counter the wind forces—in his Cincinnati and Brooklyn Bridges, but later suspension bridge designers made roadways progressively longer and less deep, thus making them more flexible not only under the traffic but also against the wind, and at the same time eliminating the diagonal cables that are almost a trademark of Roebling's bridges. Thus suspension bridges evolved through the first third of the twentieth century into longer and sleeker designs that included the George Washington Bridge, the Bronx-Whitestone Bridge, and the Tacoma Narrows Bridge. This last, of course, took slenderness beyond any reasonable limits of experience.

The paradox of engineering design is that successful structural concepts devolve into failures, while the colossal failures contribute to the evolution of innovative and inspiring structures. However, when we understand the principal objective of the design process as obviating failure, the paradox is resolved. For a failed structure provides a counterexample to a hypothesis and shows us incontrovertibly what cannot be done, while a structure that stands without incident often conceals whatever lessons or caveats it might hold for the next generation of engineers. Nowhere is the constant back-and-forth interplay between success and failure so dramatic as in the history of suspension bridges, though daring new buildings whose roofs behave not unlike bridges may hold some important lessons for the future. Othmar Ammann, who designed the George Washington and other monumental bridges, has written:

> . . . the Tacoma Narrows bridge failure has given us invaluable information. . . . It has shown [that] every new structure which projects into new fields of magnitude involves new

problems for the solution of which neither theory nor practical experience furnish an adequate guide. It is then that we must rely largely on judgment and if, as a result, errors or failures occur, we must accept them as a price for human progress.

The Tacoma Narrows Bridge was one of the most spectacular failures in the history of engineering. This first suspension bridge connecting the Olympic Peninsula with the mainland of Washington State had a narrow, two-lane center span over a half mile long. It was an unconventional design in that the depth of the roadway structure was diminished by employing a stiffened-girder design rather than the then-customary and necessarily deeper open truss. This innovation gave a slender silhouette whose appearance was dramatic and graceful, but the shallow, narrow span was also very flexible in the wind.

The roadway of the Tacoma Narrows Bridge undulated dramatically during construction and continued to behave abnormally for months after it was opened to traffic in 1940. The bridge, which came to be known as Galloping Gertie, drew thrill seekers who wanted to experience the drive-on roller coaster. On the day of its collapse the bridge's roadway gave fair warning of its final fling, and it was closed to traffic before conditions got too dangerous. The last spectacular motions of the roadway being twisted to destruction were recorded on newsreel film even as engineers were trying to understand the phenomenon of its aerodynamic instability with a scale model. The film of the last minutes of the Tacoma Narrows Bridge is a classic.

Subsequent analysis of the Tacoma Narrows failure confirmed that the bridge span acted much like an airplane wing subjected to uncontrolled turbulence. This aerodynamic aspect of bridge design was one that is no longer overlooked, and susceptible bridges contemporary with the Tacoma Narrows were quickly stiffened against crosswinds with steel that may have ruined their gracefulness but insured their structural safety. Subsequent designs were tested in wind tunnels much the way new airplane designs are.

The possibility of failure of the Tacoma Narrows Bridge in a crosswind of forty or so miles per hour was completely unforeseen by its designers, and therefore that situation was not analyzed. On paper the bridge behaved well under its own dead weight and the light traffic it was to carry. Since its use was projected to be much less than that of the similarly designed Bronx-Whitestone, the Tacoma Narrows Bridge was only two lanes wide as compared to the six-lane width of its sister bridge in New York. The Bronx-Whitestone had opened a year earlier than the Tacoma Narrows and had already had extra cables and stiffening devices installed to reduce the surprisingly large motions exhibited in the wind. These corrective measures held the up and down motion of the roadway to a matter of inches, but the undulations continued to be noticeable to the occupants of cars advancing stop and go across the bridge during rush hour. Many an afternoon, commuting home to Long Island from the Bronx, I saw and felt the bridge move through unsettlingly large distances as I waited in traffic jams caged midway between the towers and the open walls of steel that were added to stiffen the bridge after the Tacoma Narrows collapse. As late as the 1980s additional corrective measures were being taken to further reduce the motion of the bridge in the wind.

The motions of the Tacoma Narrows Bridge were measured in feet rather than inches, however, and it did not survive long enough after its opening for corrective measures to be taken. Even as it was being ripped apart by the wind, a scale model was being tested at the University of Washington to understand the phenomenon and to see what could be done. A similar model of the Bronx-Whitestone existed at Princeton University, and while both were able to reproduce the disturbing motions of the bridges, neither exactly explained them. It took the headline-grabbing story of the collapse of Galloping Gertie to bring forth the phenomenon of aerodynamic instability in suspension bridges as an explanation of how a slender bridge deck can act like an airplane wing in the wind.

Among the first hints of the true cause of the bridge's failure was a letter to *Engineering News-Record* from Theodore von Kármán, then director of the Daniel Guggenheim Aeronautical Laboratory at the California Institute of Technology. His letter, in which he employed the differential equation for an idealized bridge deck twisting like an airplane wing as the lift forces of the wind want to bend it one way and the steel of the bridge another, is a model of conciseness and what is known as back-of-the-envelope calculation. He showed that in cases where the suspension cables provide the principal resistance to twisting a simple solution to the equation can be obtained to predict the critical wind speed at which a bridge deck of narrow and shallow dimensions can be expected to twist dangerously. Von Kármán's analysis was capable of expressing the facts that the Tacoma Narrows Bridge was narrower than the Bronx-Whitestone Bridge by a factor of about three, and shallower in comparison to its own span, then the third longest in the world, by a factor of about two. Thus von Kármán could demonstrate that the Tacoma Narrows Bridge should have exhibited the phenomenon of aerodynamic instability more dramatically than any bridge extant. Indeed the bridge fell in a wind not ten miles per hour lighter than that calculated in the letter, which had appeared within two weeks after the accident.

There were not calculations but experiences a century old that should also have alerted the designers of the Tacoma Narrows to the kind of failure predicted by von Kármán's solution to a differential equation, however, and those chronicled disasters provide some of the strongest arguments that engineers should be familiar with the history of technology. If the designers of the Tacoma Narrows had known the story of the Wheeling Suspension Bridge, the longest span in the world when it was completed in 1849, they would have had no excuse for overlooking the wind as a possible cause of failure to be anticipated during design, not a problem to be dealt with after construction. After surviving for five years, the Wheeling bridge was destroyed in a storm. The technical literature did not go into details and thus deprived contemporary and later engineers—including those of the Tacoma Narrows—the lesson of the disaster, but a local reporter immortalized the bridge's last moments on May 17, 1854, in the Wheeling *Intelligencer:*

> With feelings of unutterable sorrow, we announce that the noble and world-renowned structure, the Wheeling Suspension Bridge, has been swept from its strongholds by a terrific storm, and now lies a mass of ruins. Yesterday morning thousands beheld this stupendous structure, a mighty pathway spanning the beautiful waters of the Ohio, and looked upon it as one of the proudest monuments of the enterprise of our citizens. Now, nothing remains of it but the dismantled towers looming above the sorrowful wreck that lies beneath them.
>
> About 3 o'clock yesterday we walked toward the Suspension Bridge and went upon it, as we have frequently done, enjoying the cool breeze and the undulating motion of the bridge. . . . We had been off the flooring only two minutes and were on Main Street when we saw persons running toward the river bank; we followed just in time to see the whole structure heaving and dashing with tremendous force.
>
> For a few moments we watched it with breathless anxiety, lunging like a ship in a storm; at one time it rose to nearly the height of the tower, then fell, and twisted and writhed, and was dashed almost bottom upward. At last there seemed to be a determined twist along the entire span, about one half of the flooring being nearly reversed, and down went the immense structure from its dizzy height to the stream below, with an appalling crash and roar.
>
> For a mechanical solution of the unexpected fall of this

stupendous structure, we must await further developments. We witnessed the terrific scene. The great body of the flooring and the suspenders, forming something like a basket swung between the towers, was swayed to and fro like the motion of a pendulum. Each vibration giving it increased momentum, the cables, which sustained the whole structure, were unable to resist a force operating on them in so many different directions, and were literally twisted and wrenched from their fastenings. . . .

We believe the enterprise and public spirit of our citizens will repair the loss as speedily as any community could possibly do. It is a source of gratulation that no lives were lost by the disaster.

This is as graphic a report as any of the newsreels that recorded the failure of the Tacoma Narrows Bridge almost a century later. Though no lives were lost in that accident either, and though the Narrows were bridged by a replacement span within a few years, it is not "a source of gratulation" that such a failure could occur in the middle of the twentieth century after there had been not only the example of the Wheeling Bridge but also Thomas Telford's Menai Straits suspension bridge. A gale only a month after it opened in early 1826 caused such drastic motion of the bridge that several vertical suspender rods and many floor beams of the then major 550-foot span were broken. A contemporary account of the failure is not so dramatic as the Wheeling reporter's, but it does make it clear that the Menai Straits structure failed due to undulations of its roadway. It was repaired and strengthened but suffered extreme motions and major damage in subsequent storms. Other examples from history presaged the failure of the Tacoma Narrows Bridge, but their lessons were apparently not heeded by the modern designers. And if the potential for disaster in a bridge or any structure is not caught in the blueprints, then it may take the engineers and the public as much by surprise when it happens as the behavior of the Tacoma Narrows did in 1940.

If failures remembered can be responsible for better bridges, structural successes can be responsible for better bridge builders. The Brooklyn Bridge inspired David Steinman, who grew up in its shadow, to become a builder of suspension bridges himself. As the Tacoma Narrows Bridge was twisting itself apart, he was beginning to write the story of John Roebling, his son Washington, and Washington's wife, Emily, who represented her husband at the construction site when an accident in one of the tower caissons crippled him and confined him to his room overlooking the bridge. Steinman, the designer of over four hundred bridges of his own, claimed to have taken five years out of his professional life to repay his debt of inspiration to the Brooklyn Bridge by researching and by writing the story of the Roeblings, and he was subsequently entrusted in 1948 with the task of rebuilding the historic structure to enable it to take the heavier traffic of the modern age. Thus the bridge that had almost become a monument by the late 1940s was rejuvenated to carry a full complement of six lanes of automobiles where it had once been restricted to two.

Steinman designed his own masterpieces, including the great Mackinac Bridge between the upper and lower peninsulas of Michigan. Yet, ironically, its structural success no doubt owes more to the collapse of the Tacoma Narrows than to the Brooklyn Bridge. By understanding what went wrong with Galloping Gertie, Steinman could take steps to obviate that happening in his own structure. Other bridge designers countered the lessons of the Tacoma Narrows with other solutions, however, and the Severn Bridge linking southeastern Wales and southwestern England across the estuary of the same Severn River spanned by the very first iron bridge in 1779 has turned out to be an embarrassment. Held up as an innovative solution to the problems of wind loading when its superstructure was designed in the early 1960s, the Severn Bridge has a roadway whose cross section has been described as a wing. The lightweight box girders comprising it were designed, as was the Brooklyn Bridge, to take the kinds of traffic loads projected at the time of its conception. But whereas the Brooklyn Bridge served traffic for over a half century before it had to be strengthened, the Severn Bridge began to feel the increased load of traffic within fifteen years of its opening.

The inclined hanger cables began to fray and have had to be replaced, and use of the bridge has had to be restricted to one lane of traffic in each direction. These measures have been necessary because the bridge apparently was designed with only a small margin of reserve strength, and the traffic load was a significant proportion of the bridge's weight. While the bridge sits steady to the eye in the high winds on the Severn, its cables and its roadway feel excessive load ranges that are aggravated by the corrosive action of the moist air trapped within the roadway girders and blowing through the distinctive but destructive pattern of its hanger cables, which hardly evoke the diagonal lines of the Brooklyn Bridge either structurally or aesthetically. Thus the grand successes of the Stephensons, Roeblings, and Steinmans give models and inspiration—but not rules—to their successors. The great bridges live on as symbols that innovation need not be doomed, but at the same time these symbols can be sources of false confidence.

Structural failures and surprises might all be ended if we were to stop innovating altogether. Every new bridge could be an exact copy of one that already has stood the test of time, but traffic on the new bridge could never be allowed to surpass that on the old. No new materials could be used, and no new bridge could be located on a river that did not possess the exact foundation and wind conditions of existing successful bridges. We could virtually end all risk of failure by simply declaring a moratorium on innovation, change, and progress. And it would be a moratorium on progress, for without allowing change we would in effect not allow any bridge to be built where one had not been built successfully before. For no place is quite like any other, no traffic pattern quite like any other. Even should we convince ourselves that we could reproduce a bridge with an equivalent batch of construction materials and with the equivalent quality of construction, we would halt progress of commerce in the region served by the bridge, for that would load the bridge beyond its prior experience.

Yet even the poets and painters who adopt the Brooklyn Bridges and Eiffel Towers as their artistic symbols do not expect them to remain static words on paper or paint on canvas. The bridge of Hart Crane and Joseph Stella was called upon over the years to shoulder more and more significance in their creations. As their art evolved, so did their symbols. Just as no one expects there to be a final avant-garde, so there should be no expectation that there will be a final bridge. The great bridges, like the great artistic creations of Shakespeare and Michelangelo, urge the younger generation to do more than copy the masters. While copying might be all right during one's student years, one does eventually want to create something of one's own, and that something, which might be a bridge as easily as a poem, will be an excursion into the unsure. If an inspired design relies on sound principles and does not try to extend the limits of art or engineering too far too soon, it stands a good chance of joining the canon of successes.

If structural successes egg engineers on to more daring designs and cause them to be overconfident in what appear to be designs that do not seem particularly innovative, then structural failures are reminders that they may have gone too far too soon. Thus structural failures are often the subject of extended investigations not so that we may gloat over another's mistakes, but so that all engineers may understand more fully what some apparently did not.

14

FORENSIC ENGINEERING
AND ENGINEERING FICTION

The *Alexander L. Kielland* was ten thousand tons of steel welded into a five-footed monster of an oil rig capable of virtually walking on water. As originally conceived, the ungainly vessel known as a "semi-submersible" would move from place to place in the ocean, stopping here and there to straddle geologically likely spots and drill for oil beneath the sea. In places where it might strike it rich, a permanent production platform could be erected, as if a monument to the peripatetic rig's success.

Such potential glory was never to be realized by the *Alexander Kielland,* however. In the tricky and fast-moving currents of ocean engineering, the *Kielland* was considered obsolescent by the time it was completed in 1976, and the French-built "pentagone" structure (whose five large pontoons kept the platform high and dry during transportation but were filled with water and "semi-submersed" seventy feet into the sea for better stability at anchor) was converted by its Norwegian owners into a floating hotel, or "floatel," to house as many as 348 men working on the more advanced rigs already operating in the North Sea. To add insult to injury, the *Kielland*'s 130-foot high drilling derrick, impotent though it was, was left in place on the rig's deck, surrounded not by the miles of drill pipe it may have lifted and sunk into the seabed but by scores of modular sleeping quarters and common rooms. For years, offshore crews came home to the *Kielland* and left by helicopter to work on the nearby platforms much as white-collar workers might commute by bus or train between their bedroom suburbs and a big city.

In the early evening of March 27, 1980, 212 men at home on the *Kielland* were riding out some poor weather by eating dinner, reading in their rooms, or relaxing in the sauna or the cinema. The air outside was cool but not freezing, the wind was gale- but not hurricane-force, and the waves were high but not excessively so. The Edda section of the Ekofisk oil field in the North Sea had seen a lot worse than the forty-degree Fahrenheit temperature, forty-mile-per-hour winds, and twenty-five-foot waves prevailing where the floatel was anchored at the time. The *Kielland* was designed for much more severe conditions—conditions that might be expected only once every hundred years. Thus the occupants of the rig, though perhaps uncomfortable because of the rough sea, should not have been overly concerned for their safety. Even should it suffer the calamity of having one of its five legs break away, the structure was supposed to react by listing slowly, giving the *Kielland*'s occupants plenty of time to get into life jackets and to board the lifeboats.

Unfortunately, when one of the legs did break off, with "an almighty crack," about 6:30 that evening, the platform began to quiver and the rig tilted at a steep angle much faster than predicted. The wind happened to be blowing in a direction that caught the tilting deck of the rig, which, according to some, had been made top-heavy by the dormitory modifications. The lifeboats, some of which had become overturned and tangled in their cables, could not be deployed according to plan. There were 123 victims and only 89 survivors of the offshore industry's worst accident.

Speculations about what caused the *Kielland*'s leg to break away ranged from a collision with the nearby *Edda* oil production platform or with a submarine to basic design flaws. The latter explanation seemed unlikely since the *Kielland* was among the newest of about a dozen similar rigs, and basic design flaws could be expected to show up in the oldest platforms first. Other early failure theories were also soon discounted, however, for the separated leg was towed to a calm fjord near Stavanger, Norway, where the fracture surfaces were examined. One of the braces that had broken showed the telltale signs of the uncontrollable growth of a large crack that had developed unbeknownst to the rig's operators. With each wave a crack in an offshore structure might grow less than a millionth of an inch, but since the sea will batter a structure like the *Kielland* millions of times each year, sizeable cracks can develop over time. Engineering structures can continue to function even with fatigue cracks as long as they do not reach such large proportions that they let go suddenly the way a piece of wood does under an advancing saw. Metal fatigue is designed against by eliminating all but the most minute cracks in new engineering structures, by inspecting the structure at regular intervals to detect any cracks that might be growing, and by taking the structure out of service before those minute cracks grow to dangerous proportions. (Thus an offshore oil platform might be designed to have a fatigue life of about the time it would take to extract all the oil from the location over which it sits.)

Fatigue failure can be obviated by making sure that no cracks or crack-like flaws are present in a new structure, for tens of millions of cycles of safe-life can be guaranteed to be taken up just in initiating cracks in sound metal or in growing benign cracks from acceptable imperfections. However, post-accident investigations of the *Alexander L. Kielland* revealed that when new it must have had a crack almost three inches long in a weld where an instrument bracket was attached to the ill-fated leg brace. The theretofore undetected crack was known to have existed from the moment the rig was launched because three inches of the broken crack surfaces were covered with paint that could only have been sprayed into the crevice when the final touches were being put on the new rig. The rest of the fracture surfaces showed the characteristics of fatigue crack growth up to the point where the metal finally broke apart suddenly, no longer able to hold together with such a severe flaw. One wave just slightly larger than its predecessors would have been more than the cracked structure could withstand.

After its accident, the bulk of the *Alexander L. Kielland* was towed to Stavanger Bay, where it floated upside-down for more than three years while debates about its fate took place in Norway. The uprighting of the capsized rig was described as one of the greatest feats in salvage history, and the fact that thirty-one unaccounted-for bodies might still be inside the floatel's cinema or other rooms made the ultimate disposition of the rig a very emotional issue. Demands for the recovery of the bodies led the Norwegian government to spend $34 million in a third and finally successful attempt to upright the *Kielland* with an elaborate plan involving shore-based winch chains, steel buoyancy tanks, and barges to keep everything in line. The uprighting operation was successful, but only six bodies were found on board. After the wreckage was examined by experts ranging from policemen (to investigate allegations of drug dealing) to insurance men (to collect evidence against the manufacturer), the *Alexander L. Kielland* was finally blown up and sank in 2,300 feet of water in November 1983. Though it is now out of sight, it will not be out of the minds of the relatives of the victims or the engineers who wish to design new offshore rigs and platforms with more reliability.

The investigation of the wreckage of the *Kielland* is an example of forensic engineering known as failure analysis. Although some of the acute interest in accident postmortems no doubt stems from legal and insurance claims, there is considerable engineering experience to be gained in understanding exactly what caused a structural failure. It is especially important to understand how a structure like the *Kielland* collapsed so that rational decisions can be made about whether or not to modify the design or use of other structures just like it. The grounding of the entire fleet of McDonnell-Douglas DC–10s in 1979 was a precaution taken in light of the possibility that the catastrophic Chicago crash was symptomatic of something generically wrong with the design of the plane itself. Although such curtailments of service can cause financial difficulties for the airlines and airplane manufacturer, they are necessary and necessarily drawn out because they can involve a painstaking sifting and analysis of clues as subtle as Sherlock Holmes ever had to deal with. The initial reports of failure analyses can often be premature, and the final reports coming after deliberate study can be downright wrong.

A classic example of forensic engineering involves the development of the first commercial jet aircraft. It was the first anniversary of jetliner service, May 2, 1953, when a de Havilland Comet aircraft was destroyed on taking off from Dum-Dum Airport in Calcutta, India. The fact that debris was scattered over a large area suggested that the plane disintegrated in the air, and a severe tropical thunderstorm with its attendant lightning was easily blamed for causing the tragic crash. Inspection of parts of the wreckage indicated that the plane's tail portion may have broken after being struck by something heavy, but what that might have been could not be determined. Nevertheless, this kind of observation contributed to the conclusion that a fire started *after* the plane broke up and was not the cause of the accident. The Indian Central Government Board of Inquiry concluded officially that the accident resulted from some kind of structural failure due to the plane being subjected to too much force either by severe storm gusts or by pilot overcontrol in reaction to them. Since the forces that caused the Calcutta crash were assumed to be beyond those reasonably to be expected in flight, the plane's basic design was absolved of blame. Thus the first jet passenger plane, which designers said had "flown off the drawing board" because it went into production without a prototype ever having been built, remained an engineering triumph. The Comet's designers generally believed that they had been extra-conservative in pushing aircraft technology to new heights, and they believed they had over-designed the aircraft. As long as the Comet crash could be blamed on severe weather conditions or pilot error, its design was not a prime suspect.

Thus the Comets themselves were presumed safe and returned to service without serious suspicion and without incident for another eight months. However, on January 10, 1954, a Comet taking off from Rome under mild and clear weather conditions exploded at 27,000 feet, and the pieces fell into a large circle near the island of Elba in the Mediterranean Sea. It was difficult to recover much of the plane, and the parts that were found provided no strong clues to the investigators at the Farnborough Royal Aircraft Establishment in England, where the investigation into the crash was conducted. Again there was no incontrovertible evidence that the plane's design was at fault.

Many possible causes of the second Comet crash were put forward, but no conclusion could be reached with the paucity of evidence available. Thus, although shrouded in mystery, the Comets were again put back into service within about ten weeks of the Elba crash, and on April 8, 1954, the third and final mid-air explosion of a Comet occurred as the flight en route from London to Cairo and Johannesburg was taking off from Rome. Since this wreckage fell in water too deep to leave much hope of recovery of the pieces, the search for parts of the Elba crash was renewed and intensified. Calculations were made at Farnborough, and balsa wood models of Comets were exploded over the sea to determine where the various parts would fall. Searchers thus guided and assisted by underwater television cameras recovered more of the Elba Comet and investigators continued to try to piece the clues together. Conclusive evidence that forward parts of the plane broke before the tail did was found when the tail plane was finally recovered. On inspection the tail plane was discovered to have been hit so violently by a newspaper that the reverse image of the newsprint could still be read after months of submersion in the Mediterranean. Further incontrovertible evidence that the Comet's pressurized cabin had exploded open before the tail broke away was provided by the unmistakable impression on the tail plane of an anna, an Indian coin, that no doubt was blown out of the passenger cabin back toward the tail.

Knowing that the cabin *did* explode does not explain *why* it failed. That information was provided only after an actual Comet was retired from service to be immersed in a tank of water that could be alternately pumped into and out of the cabin to simulate the repeated pressurization and depressurization that would accompany the aircraft's flight schedule. The wings were simultaneously flexed by hydraulic jacks to represent the action of the air during flight, and the test continued for about three thousand simulated flights. Suddenly, at a number of flights considerably less than the ten thousand the Comet was thought to be capable of taking before developing any significant fatigue problems, the corner of one of the cabin windows developed a crack that grew rapidly under continued flight simulations and finally shot catastrophically through the plane's metal skin. The sequence of events was arrested as soon as the water pressure was relieved by the opening, but when pressurized by air in actual flight, such a crack would be driven faster than the compressed air could escape and lead to the explosive failures theretofore unexplained. Indeed, had the testing tank experiments not been tried almost in desperation by engineers at Farnborough, the Comets may have flown again until the susceptibility of the design to fatigue was discovered by other means.

The Comet's highly pressurized cabin was known to be an important new structure when it was designed, but fatigue was not anticipated to be a problem because the pressurization and depressurization would occur slowly and only once each flight. Thus it would take an unconscionable amount of flying time to reach the hundreds of thousands or even millions of cycles that were then commonly believed to be associated with fatigue damage. (While a railroad train axle might rotate a thousand times for every mile put on the train, the early Comets that failed had accumulated only several thousand hours of their estimated lifetime of thirty thousand hours or so.) Thus the Comet designers had not worried about fatigue but about the ability of the cabin to contain the pressure itself. Although the airplane would only be pressurized to 8¼ pounds per square inch in excess of the outside pressure, the walls were designed to contain as much as twenty pounds per square inch. This corresponded to a factor of safety of almost 2½, greater than the customary factor of safety of 2 used in the aircraft industry at the time. Furthermore, the cabins of all newly manufactured Comets had been filled with compressed air and actually tested once to 16½ pounds per square inch before they were ever flown in commercial service, and this proof test was generally thought to be sufficient to exclude a fatigue failure. Indeed, the report of the findings of the Court of Inquiry into the Comet disasters states:

> Throughout the design de Havillands relied on well-established methods essentially the same as those in general use by aircraft designers. But they were going outside the range of previous experience and they decided to make thorough tests of every part of the cabin structure.
> . . . They believed, and this belief was shared by the Air Registration Board and other expert opinion, that a cabin that would survive undamaged a test to double its working pressure . . . would not fail in service under the action of fatigue. . . .

Their belief was of course wrong, and the experience with the Comets eventually improved the state of the art of aircraft design. Once the fatigue problem was understood to be aggravated by the high stresses associated with rivet holes near the window openings in the fuselage, the whole length of the window panel was replaced with one containing special reinforcement around the openings. This much increased the strength of the cabin against fatigue, and tests showed it could withstand over a hundred thousand reversals in pressure, representing hundreds of thousands of hours of flying time. The new model Comet 4 could thus conservatively be claimed safe when inspected every ten thousand hours for any telltale signs of danger.

In his autobiography, Sir Geoffrey de Havilland recounts the ordeal of the Comet, and he mentions that "many well-meaning people suggested" that the name of its successors be changed after all the Comet's early troubles. But, rather than rename the plane that had been the pride of the company and had been honestly designed to help Britain's postwar economic recovery by capturing the developing long-distance air travel business, Sir Geoffrey's firm stuck with "Comet" and only tacked a "4" onto its end to signal a new generation. He opposed a name change because, in his own words,

> . . . It seemed to me to be like a clumsy sort of cheating and would do more harm than good. It would be known sooner or later that we had deceived by a not very clever trick, and it was far better to prove that Comet failure was to be turned into Comet success.

And it was. For a while it had looked as if the Boeing Company on the other side of the ocean would be the first to put jet planes into regular passenger service across the Atlantic, but in 1958 it was a de Havilland Comet 4 that initiated trans-Atlantic jet passenger service and the plane's name was vindicated. The achievement also underscored the fact that structural failure can indeed lead ultimately to success, though sometimes the process takes a while. Technologists, like scientists, tend to hold on to their theories until incontrovertible evidence, usually in the form of failures, convinces them to accept new paradigms.

Nevil Shute Norway joined the de Havilland Aircraft Company in 1922, and, as a young aeronautical engineer fresh out of Oxford, in 1923 he began to work full time for the firm as a stress calculator. He wrote his first novel during his off hours in those early years, but he shelved it after it was rejected by several publishers. He continued to pursue writing, however, and did have a novel published in 1926. In the meantime Norway's engineering career grew, as he moved from de Havilland's to join another airship project and then went on to help found a new airplane manufacturing company, Airspeed, Ltd., in 1930. As his own growing firm took more and more of his time, he stopped writing, and he was elected a Fellow of the Royal Aeronautical Society in 1934. He resumed writing in 1937, and when the film rights to his novel *Ruined City* made him financially secure, he devoted all his time to being a writer, using the pen name of Nevil Shute. Today he is best known as the author of the story of a nuclear holocaust, *On the Beach,* which was written in the mid-1950s, and of *A Town Like Alice,* which he wrote after a trip to Australia in the late 1940s and which was dramatized on Public Television's *Masterpiece Theatre* twenty years after the writer's death in 1960.

One of Nevil Shute's less widely known novels was published in 1948 and curiously foreshadowed the elusive fatigue problem of the de Havilland Comets. The novel, entitled *No Highway,* is about a research scientist in the Structural Department of the Royal Aircraft Establishment at Farnborough. The researcher, named Theodore Honey, as a result of his fundamental studies into the fatigue of metal alloys, predicts the structural failure due to fatigue of a new trans-Atlantic airliner called the Reindeer. The plane's design is subject to the undesirable aerodynamic phenomenon of flutter, in which aggravated cycles of strain occur at high frequency and hence bring on fatigue problems much earlier than might ordinarily be expected. Honey's calculations show that the

affected tail plane assembly of the new aircraft should fracture due
to metal fatigue after only about fourteen hundred hours of flying
time, and he tries to warn the technological and technocratic
establishment of this new phenomenon in the aircraft industry. All
the Reindeers flying about the world should be checked carefully
for signs of fatigue, Honey warns. But the technocrats do not
accept his theoretical calculations as relevant to the practical matter of real airplanes, even though fatigue problems accompanied
the introduction of other new forms of transportation technology,
such as the railroads. Thus Honey sets out to collect evidence from
a Reindeer that had crashed in Canada after about the same
number of flying hours as predicted by his calculations. That crash
was attributed to simple pilot error, and hence an exhaustive
investigation of the wreckage was not considered necessary to
provide any essential evidence to the contrary. So Honey flies to
Canada himself to collect the evidence of a fatigue failure that he
is sure is lying there in the snow.

While en route, Honey discovers that the plane he is traveling
on is a Reindeer with about fourteen hundred hours of flying time,
and thus he fears that its own tail plane will fall off at any moment
due to metal fatigue. After his frustrated attempts to convince the
crew to keep the plane on the ground at Gander, a refueling stop,
until it can be properly inspected for dangerously large fatigue
cracks, Honey keeps the Reindeer from taking off by sabotaging
the landing gear. His uncommonly violent act convinces his superior to aid Honey in continuing his quest. When the wreckage in
Canada is finally reached, the telltale signs of fatigue vindicate
Theodore Honey, and a dangerous fatigue crack is also found in
the grounded Reindeer at Gander.

Nevil Shute's seemingly farfetched plot is eerily close to what
happened with the de Havilland Comets. It is as if Shute had
constructed an elaborate scenario based on Murphy's Law and
followed it to its logical conclusion. Given that new alloys behave
in new and generally unknown ways with regard to fatigue crack

growth, and given that the designers of actual jet passenger aircraft would naturally use new, lighter weight and higher strength alloys in novel ways to get their planes to fly faster, smoother, quieter, higher, and more economically than the piston-engine aircraft they would supersede, there were likely to be surprises with the revolutionary structures.

Nevil Shute appended an "author's note" to *No Highway*. It states:

> This book is a work of fiction. None of the characters are drawn from real persons. The Reindeer aircraft in my story is not based on any particular commercial aircraft, nor do the troubles from which it suffered refer to any actual events. . . .
>
> . . . The scrupulous and painstaking investigation of accidents is the key to all safety in the air, and demands the services of men of the very highest quality. If my story underlines this point, it will have served a useful purpose.

One cannot help but think that Shute is merely protesting too much in insisting on repeating the novelist's familiar white lie that any resemblance of his characters to real people is coincidental, and one wonders how much Shute believed that the development of fatigue-prone aircraft seemed almost inevitable. After all, he had not only worked as an aircraft engineer who knew the limitations of structural design and analysis but also had worked as a novelist and knew the eternal optimism of the human designer and analyst.

No Highway was published only one year after the general layout of the Comet had been finished. For some time Shute had been a close friend of Sir Alfred Pugsley, head of the Structural Department at Farnborough, where pioneering work on metal fatigue in military aircraft had been carried out in the early to mid-1940s, and thus the author was in a position to be quite

familiar with the technical details of which he wrote. Though he
may not have known for sure that it would be the Comet design
that failed, Shute recognized that there was a good chance that if
not the Comet, then the Cupid or the Donner or the Blitzen or
some other flying reindeer might easily succumb to the fatigue
problem that the arrogant and overconfident aircraft industry had
dismissed as incredible.

The single-minded purpose of Theodore Honey in rejecting the
conventional wisdom about the Reindeer accidents and the real-
life Farnborough researchers who put a Comet in a water tank are
reminders that finding the true causes of failure often take as much
of a leap of the analytical imagination as original design concepts.
And collective assent to a plausible but not incontrovertible explanation for a structural failure can allow further generic accidents
as readily as can the collective but unsubstantiated belief by a
design team that they have anticipated all possible means of failure.

Stephen Barlay, whose international documentary report on
the investigation of commercial aviation accidents was published
in Britain under the title *Aircrash Detective,* points out that even
when it seems incontrovertibly clear what is the cause of an aircraft accident, the final report only concludes a "probable cause"
because there is no knowing when future evidence, information,
or technological understanding will provide a "cause behind the
cause." He relates the story of Icarus, whose ill-fated wings represent "the first structural failure of aviation," and shows how modern knowledge of materials and structural failure can provide
alternate explanations of what happens in the myth. He then, with
complete disregard for the feelings of classicists, weaves what
might be called a technological myth for the modern age. But just
as the traditional myths of Greece and Rome have conveyed their
wisdom through hyperbole, so does this mock-heroic flight of
forensic fancy have a lesson for the engineers who would jump to
conclusions. And since myths like that of Icarus have survived

millenia of literary retelling and revising, they should emerge no less mythological from a playful engineer's analysis.

Barlay asks two questions that he claims have been enigmas for centuries. First, What was the cause behind the cause of the failure of Icarus' wings? Second, Is there a logical explanation why an island and a sea, both named after Icarus, are twenty-five miles apart? In the absence of pieces of the wreckage of the wings of Icarus, Barley goes on to examine the reports of ancient witnesses and to employ the advice of the modern Meteorological Office at Bracknell and the Accident Section of the Royal Aircraft Establishment at Farnborough.

Daedalus, it will be recalled, designed and constructed wings that he and his son, Icarus, could attach to themselves and use to fly out of the labyrinth in which they were trapped. Being the first engineer, Daedalus knew that his structure had limitations and, while the contraptions should have been sufficient to carry him and Icarus to freedom, the wings were not designed to do more than that nor to endure certain unreasonable conditions. In particular, Daedalus warned Icarus that flying too close to the sun could melt the wax affixing the feathers to the frame of the wings, and flying too close to the sea could foul the system with moisture. As Ovid reported the caveats from the mouth of Daedalus:

> "My Icarus!" he says; "I warn thee fly
> Along the middle track: nor low, nor high;
> If low, thy plumes may flag with ocean's spray;
> If high, the sun may dart his fiery ray."

Icarus, apparently being a daring young man on a flying trip, easily forgot or ignored his father's warnings and has long been believed to have flown higher and higher toward the warming yet dangerous rays of the sun to escape the chills of flight. The conventional interpretation of the myth has it that the sun did indeed melt the wings, and without his plumage Icarus plunged into the sea that bears his name.

According to Barlay, whose researchers dropped feathers to calculate their rate of descent from the mid-air separation from an errant boy's wings and studied the prevailing winds to see where the pieces of wreckage would land, the separation of the namesake island and sea could be explained rationally if Icarus were flying at an altitude of about three thousand feet. This height seems reasonable if the aviators were indeed to appear to eyewitness fishermen, as Ovid again reports, to be flying at the height of gods rather than the more common level of birds. But Barlay points out that at that altitude the air is cool rather than warm, and rather than have melted, the wax could have become brittle and broken, like the metal of a Comet or a DC–10, without warning. This indeed puts forth a new probable cause for the failure of Daedalus' structure, and it makes the myth if anything more down to earth than the traditional version. All who doubted the myth because they did not believe man could fly to the sun can now believe.

The rational analysis of failures, whether in fact, in fiction, or in myth, is of incalculable value to the engineer, for it is as much the designer's business to know how any structure might fail as it is a chess player's to know how one false move may lead to checkmate. Failures are the accidental experiments that contribute to the engineer's experience, just as the colossal mistakes of chess masters should be lessons for students of the game.

But engineering design is more complicated than a game of chess, even though Mother Nature, if she is thought of as the defending world champion, responds more predictably to our moves than does any Russian opponent. What complicates the design game is that the engineer does not always realize all the implications of the design move he himself is making. Thus, no matter how well he understands Mother Nature's strategy, he may not anticipate her response because he does not see his own move from her side of the board. Though no engineer foresaw the crack in the newly painted *Alexander L. Kielland,* the forces of Nature could not ignore it; though no engineer believed the Comet to be susceptible to fatigue, the metal could not help but crack ever so deliberately toward destruction; though no engineer anticipated the abuse a DC–10 pylon would suffer during maintenance, the structure could not hold itself together with good intentions; though no engineer redesigned the Hyatt Regency walkways to be barely able to hold themselves up, the connections could summon no more strength than they had in Kansas City.

No matter how tragic a failure might be, it is obviously more tragic if it could have been anticipated and prevented. But anticipating *all* conceivable ways a structure may fail is not as easy as looking at all possible combinations of response to a chess move. The designers of the *Alexander S. Kielland* apparently considered the loss of one of the platform's five pontoons, but they did not correctly predict what its consequences would be. Thus they thought that they had properly anticipated the precipitous collapse of the entire platform under those conditions. And if someone had conjectured that a large crack might be introduced in welding a hydrophone mounting on a leg brace, the engineers might have responded that such a crack would be noticed and reported by the painters. Had someone asked the painters if they would report such a crack, they probably would have said, "Of course!" Had someone suggested that the painters might not report a crack as large as three inches, the engineers might have said, "Enough! That is incredible. If we continue to allow for so many 'what ifs' we should have to abandon the project."

Thus it could be with the designers of the Comet, who might have read Nevil Shute's novel a year before the first test flight, four years before the Comet saw its first airliner service and six years before the first crash. Had they taken the novel literally, they might have gone back and checked the fatigue life of the tail plane. But it was not that part that was to be the Comet's downfall, and are the engineers to be faulted for not reading Shute's novel more metaphorically? They thought they *had* designed against fatigue, and the suggestion to the contrary in a piece of imaginative writing should hardly have been sufficient to remove the designers' confidence in their own thoroughness. Besides, Shute's hero claims the lifetime of the fictitious airplanes to be exactly 1,440 hours, a much too precise number for such a calculation to be taken seriously.

Failure analysis is as easy as Monday-morning quarterbacking; design is more akin to coaching. However, the design engineer must do better than any coach, for he is expected to win every game he plays. That is a tough assignment when one mistake can often mean a loss. And when defeat occurs, all one can hope is to analyze the game films and learn from the mistakes so that they are less likely to be repeated the next time out.

15

FROM SLIDE RULE TO COMPUTER: FORGETTING HOW IT USED TO BE DONE

Twenty-five years ago, the undisputed symbol of engineering was the slide rule. Engineering students, who at the time were almost all males, carried the "slip sticks" in scabbard-like cases hanging from their belts, and older engineers wore small working models as tie clips that in a pinch could be used for calculations. When I became an engineering student myself, one of my most important decisions was which slide rule to purchase. Not only was $20 a big investment in 1959, but also I was choosing an instrument that I was told I would use for the rest of my professional life; I was advised along with all the other freshmen to get right at the start a good slide rule with all the scales I would ever need. After much comparative shopping, I chose a popular Keuffel & Esser model known as the Log Log Duplex Decitrig, and for a long time it was my most prized possession. Many of my fellow students also chose K & E rules, and the company was selling them at the rate of twenty thousand per month in the 1950s.

A slide rule was indispensible for doing homework and taking tests, for all our teachers assumed that every engineering student had a slide rule and knew how to use it. If we had not learned in high school, then we quickly studied the manual folded into the box. What our engineering instructors were interested in teaching us was not all the grand things that our various models of rules could do, but their common limitations. They told us about significant digits, for most engineering instruments then had analog dials and scales from which one had to estimate numbers between the finest divisions in much the same way we have to estimate sixteenths of an inch on a yard stick or tenths of a millimeter on a meter stick. The scales on the slide rule have the same limitations, and we were expected to know that we could only report answers accurate to three significant digits from our rules, unless we were on the extreme left of the scale where finer subdivisions existed.

We often had these things inculcated in us by trial and error. If the answer to a test question required us to multiply, say, 0.346 by 0.16892 and we reported the result as 0.05844632 we would be marked for an error in significant digits, for the result of a calculation could not have a greater accuracy than the least accurately known input number. (When older engineers write 0.346, it is implied to be known only to three digits after the decimal point, otherwise it would have been written as 0.3460 or 0.34600 or to whatever decimal place the number is known.) Since no one could ever read as many digits as those in 0.05844632 from his slide rule, the closest he would be expected to get would be 0.0585. (The extra digits were a dead giveaway that the student had forgotten his slide rule and had done the multiplication longhand on some scrap of paper and, worse yet, had forgotten the significance of significant digits.) We also learned how to estimate the order of magnitude of our answers, for the slide rule could not supply the decimal point to the product of 0.346 and 0.16892, and we had to develop a feel for the fact that the answer was about 0.06 rather than 0.6 or 0.006. These requirements on our judgment made us realize two important things about engineering: first, answers are approximations and should only be reported as accurately as the input is known, and, second, magnitudes come from a feel for the problem and do not come automatically from machines or calculating contrivances.

As I progressed through engineering school with my slide rule in the early 1960s, electronic technology was being developed that was to change engineering teaching and practice. But it was not then widely known, and as late as 1967 Keuffel & Esser commissioned a study of the future that resulted in predictions of domed cities and three-dimensional television in the year 2067—but that did not predict the demise of the slide rule within five years.

In 1968, an article entitled "An Electronic Digital Slide Rule" appeared in *The Electronic Engineer*. It could dare to prophesy, "If this hand-size calculator ever becomes commercial, the conventional slide rule will become a museum piece." In the article the authors, two General Electric engineers, described a prototype that they had built with some off-the-shelf digital integrated circuits. Their "feasibility model" looked like an electric blanket control and, at $1\frac{1}{2} \times 5 \times 7$ inches, it resembled a novel in size. Yet their marvel could give four-digit answers to any four-digit multiplicands, and it could also divide and calculate square roots, exponentials, and logarithms. It had, however, one shortcoming, and the engineers made the concession that, "Since it has no decimal points, you must figure out your decimals as with a regular slide rule." As far as cost was concerned, that of course would depend upon the cost of the components, but there remained one big obstacle in 1968: "Only the digital readout still poses a problem, since at present there are no low-cost miniature devices available. But there is no question that this last barrier will soon be overcome."

They were right, of course, and within a few years Texas Instruments had developed the first truly compact, handheld, pocket-sized calculator using an electronic chip. Texas Instruments started manufacturing pocket calculators in 1972, but they were still expensive in 1973, costing about ten times as much as a top-of-the-line slide rule. However, price breakthroughs came the next year, and Commodore was advertising its model SR-1400, a "37-key advanced math, true scientific calculator" that could do everything my Log Log Duplex Decitrig could do—and more. If one knew input to ten significant digits, then this calculator could handle it.

I was teaching at the University of Texas at Austin at the time of this great calculator revolution, and there were some engineering students whose daddies did not have to wait for the pocket calculator to become competitive in price with the slide rule. We faculty were thus faced with the question of whether students with electronic slide rules had an unfair advantage on quizzes and examinations over those with the traditional slip sticks, for the modern electronic device was a lot quicker and could add and subtract—something a slide rule could only do with logarithms. The faculty members generally were unfamiliar with all the features of the calculators that were still priced out of their reach, and there seemed to be many pros and cons and endless discussions over the issue of whether an electronic slide rule was equivalent to a wooden one. The question soon became moot, however, as prices plunged and just about anyone who could afford a conventional slide rule could afford an electronic model. By 1976 Keuffel & Esser was selling calculators made by Texas Instruments faster than traditional slide rules, which by then made up only five percent of K & E's sales, and the company consigned to the Smithsonian Institution the machine it once used to carve the scales into its wooden slide rules.

By the mid-1970s calculator manufacturers were making fifty million units a year, and soon just about everyone, including engineers who went through school in the old days, had a calculator. But no older engineer that I know discarded or consigned his slide rule to any museum. At most the old slip stick was put in the desk drawer, ready for use during power failures or other emergencies. A study conducted by the Futures Group in the early 1980s found that most engineers in senior management positions continued to keep slide rules close at hand and still used them "because they are more comfortable." But the always-growing younger generations naturally feel just the opposite. In 1981 I asked a class of sophomore engineering students how many used a slide rule, and I got the expected answer—none. (Some did own slide rules, perhaps because their engineer fathers bought one for the freshman to take away to engineering school. And K & E was selling out its remaining stock of 2,300 at the rate of only two hundred per month in 1981.) I did not ask my class how many used a calculator, for that would be like asking how many use a telephone. And I did not ask how many used a computer, for that was by then a requirement in the engineering curriculum. The trend is clearly that eventually no engineer will own or use a traditional slide rule, but that practicing engineers of all generations will use—and misuse—computers.

Engineering faculty members, like just about everyone else, got so distracted by the new electronic technology during the 1970s that more substantial issues than price, convenience, and speed of computation did not come to the fore. The vast majority of faculty members did not ask where all those digits the calculators could display were going to come from or go to; they did not ask if the students were going to continue to appreciate the approximate nature of engineering answers, and they did not ask whether students would lose their feel for the decimal point if the calculator handled it all the time. Now, a decade after the calculator displaced the slide rule, we are beginning to ask these questions, but we are asking them not about the calculator but about the personal computer. And the reason these questions are being asked is that the assimilation of the calculator and the computer is virtually complete with the newer generations of engineers now leaving school, and the bad effects are beginning to surface. Some structural failures have been attributed to the use and misuse of the computer, and not only by recent graduates, and there is a real concern that its growing power and use will lead to other failures.

The computer enables engineers to make more calculations more quickly than was conceivable with either the slide rule or the calculator, hence the computer can be programmed to attack problems in structural analysis that would never have been attempted in the pre-computer days. If one wished to design a complicated structure of many parts, for example, one might first have made educated guesses about the sizes of the various members and then calculated the stresses in them. If these stresses were too high, then the design had to be beefed up where it was overstressed; if some calculated stresses were too low, then those understressed parts of the structure could be made smaller, thus saving weight and money. However, each revision of one part of the structure could affect the stresses in every other part. If that were the case, the entire stress analysis would have to be repeated. Clearly, in the days of manual calculation with a slide rule—wooden or electronic—such a process would be limited by the sheer time it would consume, and structures would be generally overdesigned from the start and built that way. Furthermore, excessively complex structures were eschewed by designers because the original sizing of members might be too difficult to even guess at, and calculations required to assure the safety of the structure were simply not reasonable to perform. Hence engineers generally stuck with designing structures that they understood well enough from the very start of the design process.

Now, the computer not only can perform millions of simple, repetitive calculations automatically in reasonable amounts of time but also can be used to analyze structures that engineers of the slide rule era found too complex. The computer can be used to analyze these structures through special software packages, claimed to be quite versatile by their developers, and the computer can be instructed to calculate the sizes of the various components of the structure so that it has minimum weight since the maximum stresses are acting in every part of it. That is called optimization. But should there be an oversimplification or an outright error in translating the designer's structural concept to the numerical model that will be analyzed through the automatic and unthinking calculations of the computer, then the results of the computer analysis might have very little relation to reality. And since the engineer himself presumably has no feel for the structure he is designing, he is not likely to notice anything suspicious about any numbers the computer produces for the design.

The electronic brain is sometimes promoted from computer or clerk at least to assistant engineer in the design office. Computer-aided design (known by its curiously uncomplimentary acronym CAD) is touted by many a computer manufacturer and many a computer scientist-engineer as the way of the future. But thus far the computer has been as much an agent of unsafe design as it has been a super brain that can tackle problems heretofore too complicated for the pencil-and-paper calculations of a human engineer. The illusion of its power over complexity has led to more and more of a dependence on the computer to solve problems eschewed by engineers with a more realistic sense of their own limitations than the computer can have of its own.

What is commonly overlooked in using the computer is the fact that the central goal of design is still to obviate failure, and thus it is critical to identify exactly *how* a structure may fail. The computer cannot do this by itself, although there are attempts to incorporate artificial intelligence into the machine to make it an "expert system," and one might dream that the ultimate in CAD is to have the computer learn from the experience contained in files of failures (stored in computers). However, until such a far-fetched notion becomes reality, the engineer who employs the computer in design must still ask the crucial questions: Will this improperly welded pipe break if an earthquake hits the nuclear reactor plant? Will this automobile body crumple in this manner when it strikes a wall at ten miles per hour? Will any one of the tens of thousands of metal rods supporting this roof break under heavy snow and cause it to fall into the crowded arena?

One *can* ask of the computer model questions such as these. Whether or not they *are* asked can depend on the same human judgment that dismissed the question of fatigue in the Comets and that apparently did not check the effects of the design change on the Hyatt Regency walkways. Even if one thinks of the critical questions and can phrase them so that the computer model is capable of producing answers to them, there may have to be a human decision made as to how exhaustive one can be in one's interrogation of the computer. While the computer works very quickly as a file clerk, it cannot work very quickly when it is asked to analyze certain engineering problems. One of the most important problems in design is the behavior of metal under loads that deform structural components permanently. While it takes only seconds to put a bar of ductile steel in a testing machine and pull the bar until it stretches out and breaks like a piece of taffy, simulating such an elementary physical test on the largest computer can take hours.

There can be miles of pipes in a typical nuclear reactor plant, and it could take some of the largest and fastest computers a full day of nonstop calculation to determine how wide and how long a crack in one ten-foot segment of the piping would grow under the force of escaping water and steam. The results of such a calculation are important not only to establish how large a leak might develop in the pipe but also to determine whether or not the pipe might break completely under the conditions postulated (by the human engineer). Since it could take years of nonstop computing and millions of dollars to examine every conceivable location, size, and type of crack in every conceivable piece of pipe, the human engineer must make a judgment just as in the old days as to which is the most likely situation to occur and which is the most likely way in which the pipe can fail. The computer does not work with ideas but with numbers, and it can only solve a single problem at a time. The pipe it looks at must have a specified diameter, a specified crack, a specified strength, and a specified load applied to it. Furthermore, the computer model of the cracked pipe must have a specified idea as to how a crack grows as the postulated accident progresses. All these specifications are made by human beings, and thus the results of the computer are only as conclusive about the safety of the system as the questions asked are the critical ones.

The computer is both blessing and curse for it makes possible calculations once beyond the reach of human endurance while at the same time also making them virtually beyond the hope of human verification. Contemporaneous explanations of what was going on during the accident at Three Mile Island were as change-able as weather forecasts, and even as the accident was in progress, computer models of the plant were being examined to try to figure it out.

Unfortunately, nuclear plants and other complex structures cannot be designed without the aid of computers and the complex programs that work the problems assigned them. This leads to not a little confusion when an error is discovered, usually by serendip-ity, in a program that had long since been used to establish the safety of a plant operating at full power. The analysis of the many piping systems in nuclear plants seems to be especially prone to gremlins, and one computer program used for calculating the stresses in pipes was reportedly using the wrong value for pi, the ratio of the circumference to the diameter of a circle that even a high school geometry student like my daughter will proudly recite to more decimal places than the computer stores. Another inci-dent with a piping program occurred several years ago when an incorrect sign was discovered in one of the instructions to the computer. Stresses that should have been added were subtracted by the computer, thus leading it to report values that were lower than they would have been during an earthquake. Since the com-puter results had been employed to declare several nuclear plants earthquake-proof, all those plants had to be rechecked with the corrected computer program. This took months to do and several of the plants were threatened with being shut down by the Nuclear Regulatory Commission if they could not demonstrate their safety within a reasonable amount of time.

Even if a computer program is not in error, it can be improperly employed. The two and a half acres of roof covering the Hartford Civic Center collapsed under snow and ice in January 1978, only hours after several thousand fans had filed out following a basketball game. The roof was of a space-frame design, which means that it was supported by a three-dimensional arrangement of metal rods interconnected into a regular pattern of triangles and squares. Most of the rods were thirty feet long, and as many as eight rods had to be connected together at their ends. The lengthy calculations required to ensure that no single rod would have to carry more load than it could handle might have kept earlier engineers from attempting such a structure or, if they were to have designed it, they might have beefed it up to the point where it was overly safe or to where its own weight made it prohibitively expensive to build. The computer can be used to calculate virtually all the possibilities, which, so long as calculations are not made for rods that stretch or bend permanently, is not nearly so time consuming as the calculation for a cracked pipe, and engineers can gain an unwarranted confidence in the validity of the resulting numbers. But the numbers actually represent the solution to the problem of the space-frame model in the computer and not that of the actual one under ice and snow. In particular, the computer model could have understated the weight on the roof or oversimplified the means by which the rods are interconnected. The means of connection is a detail of the design that is much more difficult to incorporate into a computer model than the lengths and strengths of the rods, yet it is precisely the detail that can transmit critical forces to the physical rods and cause them to bend out of shape.

In reanalyzing the Hartford Civic Center's structure after the collapse, investigators found that the principal cause of failure was inadequate bracing in the thirty-foot-long bars comprising the top of the space truss. These bars were being bent, and the one most severely bent relative to its strength folded under the exceptional load of snow and ice. When one bar bent, it could no longer function as it was designed to, and its share of the roof load was shifted to adjacent bars. Thus a chain reaction was set up and the entire frame quickly collapsed. The computer provided the answer to the question of how the accident happened because it was asked the right question explicitly and was provided with a model that could answer that question. Apparently, the original designers were so confident of their own oversimplified computer model (and that they had asked of it the proper questions) that when workmen questioned the large sag noticed in the new roof they were assured that it was behaving as it was supposed to.

Because the computer can make so many calculations so quickly, there is a tendency now to use it to design structures in which *every* part is of minimum weight and strength, thereby producing the most economical structure. This degree of optimization was not practical to expect when hand calculations were the norm, and designers generally settled for an admittedly over-designed and thus a somewhat extravagant, if probably extra-safe, structure. However, by making every part as light and as highly stressed as possible, within applicable building code and factor of safety requirements, there is little room for error—in the computer's calculations, in the parts manufacturers' products, or in the construction workers' execution of the design. Thus computer-optimized structures may be marginally or least-safe designs, as the Hartford Civic Center roof proved to be.

The Electric Power Research Institute has been sponsoring a program to test the ability of structural analysis computer software to predict the behavior of large transmission towers, whose design poses problems not unlike a three-dimensional space-frame roof. A full-size giant tower has been constructed at the Transmission Line Mechanical Research Facility in Haslet, Texas, and the actual structure can be subjected to carefully controlled loads as the reaction of its various members is recorded. The results of such real-world tests were compared with computer predictions of the tower's behavior, and the computer software did not fare too well.

Computer predictions of structural behavior were within only sixty percent of the actual measured values only ninety-five percent of the time, while designers using the software generally expect an accuracy of at least twenty percent ninety-five percent of the time. Clearly, a tower designed with such uncertain software could be as unpredictable as the Hartford Civic Center roof. It is only the factor of safety that is applied to transmission towers that appears to have prevented any number of them from collapsing across the countryside.

In the absence of these disturbing tests, the success of towers designed by computer might have been used to argue that the factor of safety should be lowered. Conservative opposition to lowering a factor of safety would be hard to maintain for structures that had been experiencing no failures, and time, if nothing else, would wear down the opponents. But a lower factor of safety would invariably lead to a failure, which in turn would lead to the realization that the computer software was not analyzing the structure as accurately as was thought. But it would have been learning a lesson the hard way.

Thus, while the computer can be an almost indispensable partner in the design process, it can also be a source of overconfidence on the part of its human bosses. When used to crunch numbers for large but not especially innovative designs, the computer is not likely to mislead the experienced designer because he knows, from his and others' experience with similar structures, what questions to ask. If such structures have failed he will be particularly alert to the possibility of similar modes of failure in his structure. However, when the computer is relied upon for the design of innovative structures for which there is little experience of success, let alone failure, then it is as likely, perhaps more likely, for the computer to be mistaken as it was for a human engineer in the days of the slide rule. And as more complex structures are designed *because* it is believed that the computer can do what man cannot, then there is indeed an increased likelihood that structures will fail, for the further we stray from experience the less likely we are to think of all the right questions.

It is not only large computers that are cause for concern, and some critics have expressed the fear that a greater danger lies in the growing use of microcomputers. Since these machines and a plethora of software for them are so readily available and so inexpensive, there is concern that engineers will take on jobs that are at best on the fringes of their expertise. And being inexperienced in an area, they are less likely to be critical of a computer-generated design that would make no sense to an older engineer who would have developed a feel for the structure through the many calculations he had performed on his slide rule.

In his keynote address on the structural design process before the Twelfth Congress of the International Association for Bridge and Structural Engineering held in Vancouver in 1984, James G. MacGregor, chairman of the Canadian Concrete Code Committee, expressed concern about the role of computers in structural design practice because "changes have occurred so rapidly that the profession has yet to assess and allow for the implications of these changes." He went on to discuss the creation of the software that will be used for design:

> Because structural analysis and detailing programs are complex, the profession as a whole will use programs written by a few. These few will come from the ranks of the structural "analysts" . . . and not from the structural "designers." Generally speaking, their design and construction-site experience and background will tend to be limited. It is difficult to envision a mechanism for ensuring that the products of such a person will display the experience and intuition of a competent designer.
>
> In the design office the reduction in computation time will free the engineer to spend more time in creative thought—*or* it will allow him to complete more work with less creative

thought than today. Because the computer analysis is available it will be used. Because the answers are so precise there is a tendency to believe them implicitly. The increased volume of numerical work can become a substitute for assessing the true structural action of the building as a whole. Thus, the use of computers in design must be policed by knowledgeable and experienced designers who can rapidly evaluate the value of an answer and the practicality of a detail. More than ever before, the challenge to the profession and to educators is to develop designers who will be able to stand up to and reject or modify the results of a computer aided analysis and design.

The American Society of Civil Engineers considered the problem of "computer-extended expertise" such an important issue that it made it the subject of its 1984 Mead Prize competition for the best paper on the topic "Should the Computer be Registered?" The title is an allusion to the requirement that engineers be registered by state boards before they can be in charge of the design of structures whose failure could endanger life. Professional engineering licenses come only after a minimum period of engineering work with lesser responsibility and after passing a comprehensive examination in the area of one's expertise. Computers, while really no more than elaborate electronic slide rules and computation pads, enable anyone, professional engineer or not, to come up with a design for anything from a building to a sewer system that looks mighty impressive to the untrained eye. The announcement for the Mead Prize summarized the issue succinctly:

> Civil engineers have turned to the computer for increased speed, accuracy and productivity. However, do engineers run the risk of compromising the safety and welfare of the public? Many have predicted that the engineering failures of the future will be attributed to the use or misuse of comput-

ers. Is it becoming easy to take on design work outside of the engineer's area of expertise simply because a software package is available? How can civil engineers guarantee the accuracy of the computer program and that the engineer is qualified to use it properly?

By throwing such questions out to its Associate Members, those generally young in experience if not in age and the only ones eligible to compete for the Mead Prize, the ASCE at the same time acknowledged and called to the attention of future professional engineers one of the most significant developments in the history of structural engineering.

16
CONNOISSEURS OF CHAOS

A. A violent order is disorder; and
B. A great disorder is an order. These
Two things are one. . . .
—Wallace Stevens

The causes of failures can be as many and as muddled as their lessons. When something goes wrong with a computer program or an engineering structure, the scrutiny under which the ill-fated object comes often uncovers a host of other innocuous bugs and faults that might have gone forever unnoticed had the accident not happened. Had the hanger rod-box beam suspension detail of the Hyatt Regency walkways not been changed from the original concept, they would no doubt be standing today, the site of many a party and probably unsuspected of being in violation of the building code. And after the second de Havilland Comet exploded, about fifty major and minor modifications to the design were made because that many subcritical deficiencies had been identified in looking for the critical fault. However, the plethora of faults that emerges from an accident postmortem is usually dominated by a single, most probable cause. There have been numerous attempts to classify these ultimate causes of structural failures, but no two lists seem to agree as to what the categories should be.

Thomas McKaig's 1962 book *Building Failures* is a widely known collection of case studies intended for the use of engineers, architects, and contractors. Although he clearly places the blame on human error, McKaig believes that those involved in building failures are themselves the victims of accidents, and he calls for sympathy and commiseration. However, when he attempts to explain why accidents happen, McKaig states succinctly that, "Usually buildings fail through men's ignorance, carelessness, or greed." Then he goes on to quote a list of "causes for failure" that he found in a magazine "so long ago that its source is unknown." But McKaig clearly subscribes to the anonymous list, and he believes that it applies to any type of engineering construction. Furthermore, according to McKaig, "It is difficult to conceive of any failure or difficulty— major or minor—that does not fit into one of these classifications. . . .":

1. Ignorance
   a. Incompetent men in charge of design, construction, or inspection.
   b. Supervision and maintenance by men without necessary intelligence.
   c. Assumption of vital responsibility by men without necessary intelligence.
   d. Competition without supervision.
   e. Lack of precedent.
   f. Lack of sufficient preliminary information.
2. Economy
   a. In first cost.
   b. In maintenance.
3. Lapses, or carelessness
   a. An engineer or architect, otherwise careful and competent, shows negligence in some certain part of the work.
   b. A contractor or superintendent takes a chance, knowing he is taking it.
   c. Lack of proper coordination in production of plans.
4. Unusual occurrences—earthquakes, extreme storms, fires, and the like.

While the last category, with its catchall "and the like," might be considered equivalent to a "none-of-the-above" answer to a multiple-choice question, the list endorsed by McKaig is indeed a compelling one. However, under which category a given accident might be placed could easily lead to arguments in judgment, opinion, and semantics, and there is a curious distinction made between the contractor who "takes a chance" and the engineer who "shows negligence." In fact, design engineers might also be said to take chances, especially when producing a prototype whose function it is to reveal weaknesses in design concepts. But McKaig's difficulty in conceiving of any failure that does not fall into one of the categories might be said to hint at the single cause of *all* accidents. His claim for the completeness of the classification of the causes of failures is not unlike a designer's implicit claim for the completeness of his *anticipation* of all possible ways in which his structure can fail. And when the designer believes he has obviated all the potential failures on *his* complete list, then he believes his design job to be done and done responsibly. But the designer, no less than McKaig, is taking a chance on his perspicacity and prescience.

One recently published textbook on the failure of materials in mechanical design acknowledges in its preface that the concept of failure is central to the process of engineering design:

> Recognition of the potential for failure and identification of the modes of mechanical failure that persist in the real engineering world are absolutely essential to prediction and prevention of mechanical failure, the cornerstone objectives of every mechanical designer. Thus the book identifies the modes of mechanical failure early in the presentation. . . .

The book proper lists no fewer than twenty-three main modes of mechanical failure, with many subdivisions, and suggests that the list includes all "commonly observed" modes. Advertising copy, with its usual hyperbole, headed, "How to Make Your Mechanical Designs Failure-proof," claims it to be "the only available book with a comprehensive list of all mechanical failure modes and detailed definitions." But a reader may well wonder which uncommon modes will account for the unexpected failures of the new materials and designs being developed even as the textbook is being used.

Not everyone who attempts to explain structural failures claims to be able to make complete lists of causes. D. I. Blockley, whose otherwise ambitious book, *The Nature of Structural Design and Safety,* looks at the problem from several perspectives, including philosophical, historical, and analytical ones, makes no claims for completeness in his table of categories:

SOME CAUSES OF STRUCTURAL FAILURE
Limit states

Overload:	geophysical, dead, wind, earthquake, etc.; manmade, imposed, etc.
Understrength:	structure, materials instability
Movement:	foundation settlement, creep, shrinkage, etc.
Deterioration:	cracking, fatigue, corrosion, erosion, etc.

Random hazards
Fire
Floods

Explosions:	accidental, sabotage

Earthquake
Vehicle impact
Human-based errors

Design error:	mistake, misunderstanding of structural behavior
Construction error:	mistake, bad practice, poor communications

In this list the human element seems to be separated from the physical behavior of the environment in which the structure is placed. While it must be implicitly understood that it is the human designer's responsibility to anticipate what might overload his structure or what might cause it to lack strength or what might cause undesirable movement or deterioration in the structure, any real or imagined claims to completeness hinge upon the ubiquitous "etc." in the list of limit states that the designer must anticipate. How well he completes the categories and ensures his design structurally against failure due to any of the listed or unlisted problem areas will determine the safety of his structure.

The lists of McKaig and Blockley are almost diametrically opposed in their view of the role of the individual in the causes of failure. The people that dominate McKaig's categories are not even explicitly mentioned in the "human-based errors" that seem almost tacked on to Blockley's list. And the various kinds of physical forces and conditions that Blockley seems to consider as the "causes" of failures are nowhere made the explicit culprit in McKaig's anthropocentric list.

But even if a common list could be agreed upon, investigators of structural accidents might not always agree on the category in which to put a certain incident. Would the Hyatt Regency accident be listed under "building code violation" or "faulty detail"? Would the Comet crashes be blamed on "fatigue-prone material" or "insufficient analysis"? And would the 1979 American Airlines DC–10 crash in Chicago be attributed to "improper maintenance" or to an "inadequate design" that did not anticipate the improper maintenance procedures? Or would any one of these infamous failures be classified under "overloaded" or "understrength" structure? Some engineers would say it is all a matter of semantics and that all structural failures can be traced back to one cause, *design error,* for even so-called construction errors should be anticipated by the designer. It is true, of course, that all failures can be argued to be the result of design errors, for as the purpose of design is to obviate failure, the failure not anticipated is a clear indication of improper design. But to obviate failure, a designer must anticipate it.

A subcommittee of the U. S. House of Representatives Committee on Science and Technology held hearings in 1982 to examine the problem of structural failures in this country. While the subcommittee sought to identify factors that contribute most significantly to the occurrence of structural failure, its report lists not the causes of failures but several lists of significant factors that are important in *preventing* structural failures. The findings of the committee include these six "critical" factors in preventing structural accidents from happening:

1. communications and organization in the construction industry;
2. inspection of construction by the structural engineer;
3. general quality of design;
4. structural connection design details and shop drawings;
5. selection of architects and engineers;
6. timely dissemination of technical data.

Among some of the "moderately" significant factors are cost cutting on design and construction, and among the "least" significant factors are the adequacy of building codes and the impact of fast-track scheduling. Perhaps not least, though it is last on the least list, is the "need for legislative changes." Among the recommendations of the committee with regard to the timely dissemination of technical data is that the National Bureau of Standards be authorized to undertake on its own initiative the investigation of major structural failures in public structures and to collect and disseminate relevant information.

Historically, the National Bureau of Standards has conducted investigations of major failures only when requested to do so by the local authorities, who control subpoena powers to gain access to the accident site. There were some problems with investigators getting into the Hyatt Regency after its skywalks collapsed, and the NBS team could inspect the fallen structure only after the mayor of Kansas City intervened. Representative Albert Gore of Tennessee, chairman of the Investigations and Oversight Committee of the House Science and Technology Committee, would like to empower the NBS to investigate structural failures and disseminate information in much the same way the National Transportation Safety Board now is empowered to investigate aircraft and other accidents under its jurisdiction.

The committee's recommendation was motivated in part to ensure that information on structural failures is not sealed in voluminous court records following long legal battles, as was the case when the John Hancock Building in Boston suffered a chronic problem of having its windows fall out. The case for disclosure was made very convincingly by Barry LePatner, an attorney who represents many architectural and engineering firms, and a witness before the House subcommittee:

> Good judgment is usually the result of experience. And experience is frequently the result of bad judgment. But to learn from the experience of others requires those who have the experience to share the knowledge with those who follow.

The dissemination of failure information has already begun, however, and it is based on the premise that a designer can learn from the mistakes of others. In order to keep a designer from forgetting a way in which another structure similar to the one on his drawing board or computer screen may have failed in the past, an archive of structural failures was recently begun at the Architecture and Engineering Performance Information Center (AEPIC) established at the University of Maryland. This clearinghouse brings to fruition an idea that had long been promoted within the American Society of Civil Engineers, and a modest amount of funding from the National Science Foundation provided the seed money. AEPIC will keep computerized files not only on structural collapses, but also on lesser failures such as water damage due to poor roof design and architectural facades that crack or fall off. An engineer or architect intending to use a structural design or a facing material that he is unfamiliar with may call up from the computer's files all relevant experience with the intended design or material. But AEPIC will be subject to the same limitations as have been other list-makers, and its ultimate usefulness will depend to a large extent on how complete and how valid its categorization of structural failures will be and on how much in concert its point of view is with its potential users.

Whether or not the AEPIC concept works to improve the reliability of design and prevent the repetition of errors will also depend very strongly on how much cooperation the center gets from those who have records of experience for the computer files. In the impassioned preface to *Building Failures,* McKaig lamented:

> . . . It is regrettable that the legal processes of determining liability for a failure so often serve to bury information of great potential value. The insurance carriers may be satisfied with a quiet settlement behind closed doors, but I strongly believe there is also a responsibility, in these cases, to the public.

Insurance company files can indeed be especially rich stores of information for a failure archive, and the delivery of the records of forty thousand cases from one insurance underwriter in 1982 gave the fledgling AEPIC an encouraging start. However, since there can be so many claims and counterclaims associated with an engineering or architectural failure, and since engineering and architectural firms that have learned the hard way are not necessarily proud of their (sometimes dumb) mistakes, whether other firms follow with more troves for the failure archive remains to be seen. And even if the data are gathered, how they are or are not used will no doubt depend in part on how successful AEPIC is in categorizing and retrieving the data.

The timely dissemination of information on structural failures has long existed in the pages of *Engineering News-Record,* which is proud of its 110-year tradition of reporting and recording. While no one claims the back issues of *ENR* to be any official archive of failures, they certainly are a treasure trove, and the recently published book *Construction Disasters* by Steven S. Ross is an attempt to categorize some of the contents of that trove and to draw some lessons out of it.

No individual's list of the causes of failures or choices of case studies or of categories in which to put them or of lessons to be drawn from them is likely to satisfy everyone, and hence all such attempts are likely to be doomed to failure themselves. However, there is another, technologically unorthodox method of expounding on engineering design and structural failure that has the advantage of being at the same time less precise and more thought provoking. That is the method of creative writing, in which the plot of a novel or a narrative poem is constructed around a technical idea. Superficially the story can be entertaining while its message or moral can have profound implications. Furthermore, since fiction and poetry are open to interpretation, each reader can bring his own experiences and take away his own wisdom. And if the technological ideas are correct and consistent, the technical community will sit up and read.

Oliver Wendell Holmes' "The Deacon's Masterpiece," about the absurdity of building a one-hoss shay that will not break down, is one such work. Nevil Shute's *No Highway,* about metal fatigue in a new airplane design, is another. And Robert Byrne's recent novel *Skyscraper,* by having technology as its main theme, is still another example in the tradition of Holmes and Shute. The hero of *Skyscraper* is an engineer who specializes in failure analysis, and he is called to New York City to look into why a two-hundred-pound pane of glass fell out of the fictitious sixty-six-story Zalian Building. In the process of his investigation he discovers numerous flaws and danger signals in the structure, and Byrne uses this framework to explain how skyscrapers are built and how they can fail, especially how they can topple over, as the Zalian Building does in the final chapters of the suspenseful book. Although it has the requisite sex and violence of a popular novel, *Skyscraper* also has a considerable amount of technical detail and speculation, which Byrne is able to weave convincingly into the plot no doubt because of his education as a civil engineer and his fifteen years' experience as an editor of a construction industry trade journal.

While *Skyscraper* can be read as a suspense novel by the layman, who will incidentally pick up some understanding of engineering design and failure analysis, it can also be read by professionals as a hypothetical case study of a major structural failure. Byrne complicates his plot with a variety of compounding causes of structural failure, including design error, lack of objective inspection during and after construction, cost-cutting, building code violations, and the use of computers to design a lighter and more flexible structure than would have been attempted in the pre-computer era. It is as if Byrne were attempting to include in his novel as many of the conceivably possible factors contributing to structural failure that were identified by the Congressional subcommittee, whose report was issued only three months before the publication of the novel. This all suggests that the causes of structural failure are generally known and agreed upon, but categorizing and preventing them is another matter. As Byrne shows in *Skyscraper,* the human element, not only in making honest mistakes but also in condoning cover-ups, could be at the heart of the matter in at least some structural failures. It is not that anyone, honest or dishonest, really wants a building to fall down, it is that there seems often to be a general disbelief that it could really happen because of this or that particular defect.

The significance for the technical community of a novel like *Skyscraper* is in its attempt to deal explicitly with the human element. Books of case studies and lists of causes of failures do not easily incorporate this synergistic element, yet the motives and weaknesses of individuals must ultimately be taken into account in any realistic attempt to protect society from the possibilities of major structural collapses. This protection may ultimately have to come from systems of checks and balances in the construction industry, from investigative teams legislatively empowered to explicate the causes of failures that do occur, and from the maintenance of adequate factors of safety to ensure that designs are not just marginally sound. Books like Byrne's, by pointing out what can go wrong, are a positive contribution to structural safety, and they provide further evidence for the desirability of disseminating information about failures. If only one design engineer or one construction manager sees in a poem or story a single analogy that enables him to catch a flaw in his own project, the piece of imaginative writing has contributed to structural safety as surely as do building codes, failure archives, legislative hearings, or even legislation.

It is the engineer's constant challenge to conceive the new from the old, and it is his lot to worry about his curious kind of time travel that transcends the instruments of calculation and forces him always to think about the future to avoid the failures of the past. Bugs entered engineering calculations long before the computer, and wondering whether he had thought of every possible failure mode has kept many an engineer awake. Both the appeal and anxiety of engineering are evident in Herbert Hoover's reminiscence about his career as a mining engineer prior to entering politics:

It is a great profession. There is the fascination of watching a figment of the imagination emerge through the aid of science to a plan on paper. Then it moves to realization in stone or metal or energy. Then it brings jobs and homes to men. Then it elevates the standards of living and adds to the comforts of life. That is the engineer's high privilege.

The great liability of the engineer compared to men of other professions is that his works are out in the open where all can see them. His acts, step by step, are in hard substance. He cannot bury his mistakes in the grave like the doctors. He cannot argue them into thin air or blame the judge like the lawyers. He cannot, like the architects, cover his failures with trees and vines. He cannot, like the politicians, screen his shortcomings by blaming his opponents and hope that the people will forget. The engineer simply cannot deny that he did it. If his works do not work, he is damned. That is the phantasmagoria that haunts his nights and dogs his days. He comes from the job at the end of the day resolved to calculate it again. He wakes in the night in a cold sweat and puts something on paper that looks silly in the morning. All day he shivers at the thought of the bugs which will inevitably appear to jolt its smooth consummation.

17
THE LIMITS OF DESIGN

Daedalus, whose mythical wing-making has earned him the title of first aeronautical engineer, is said to have cursed his skill when he spied the wings of his son Icarus floating on the sea. But Icarus had to share some of the blame for what may well have been the first structural failure in the history of air travel. He had been warned by his father, the designer of the wax and feather wings, that he should not fly too high with the untested new invention.

Instead of wandering about the labyrinth dreaming of the perfect and indestructible wing, Daedalus accepted the compromise of a design that could be fouled by the water or melted by the sun. Everyone and everything has its limitations and its breaking point, but that does not mean that we and all our designs are total failures. Just as people are not expected to push themselves too hard or to overextend themselves, so machines and structures are not expected by their designers to be pushed too hard or to be overloaded. Daedalus foresaw, as all engineers must foresee, the ways in which his structure could fail, for it is only by recognizing the possible ways of failure that a successful structure can be designed to resist the forces that might tear part from part. Had Icarus used the wings within the proper altitude, they would to this day be declared a success and the first instruments of successful, albeit mythical, manned flight.

What Daedalus did was presumably the best he could with the technology and resources available to him. He had limited materials, so wax and feathers had to do. Neither did he and Icarus have limitless time, for the hungry Minotaur was also in the labyrinth. The tragedy of the myth of the first flight is that the flight might have worked—within the myth at least—had the repeated cautions of Daedalus been heeded by his son.

Even today, when engineers make wings of metal for airplanes that carry myriad strangers as well as their own sons, the designers still must caution the pilots and maintenance men about the proper use of the wings. This is done through operating and maintenance manuals, which provide the proper limitations and procedures, but as accidents attributed to pilot error and improper maintenance procedures testify, there is still a bit of Icarus about the airport.

The object of engineering design is to obviate failure, but the truly fail-proof design is chimerical. The ways in which a structure or machine can fail are many, and their effects range from blemishes to catastrophes. The designer David Pye has written of the compromise of design:

> The requirements for design conflict and cannot be reconciled. All designs for devices are in some degree failures, either because they flout one or another of the requirements or because they are compromises, and compromise implies a degree of failure.
>
> Failure is inherent in all useful design not only because all requirements of economy derive from insatiable wishes, but more immediately because certain quite specific conflicts are inevitable once requirements for economy are admitted; and conflicts even among the requirements of use are not unknown.
>
> It follows that all designs for use are arbitrary. The designer or his client has to choose in what degree and where

there shall be failure. Thus the shape of all designed things is the product of arbitrary choice. If you vary the terms of your compromise—say, more speed, more heat, less safety, more discomfort, lower first cost—then you vary the shape of the thing designed. It is quite impossible for any design to be "the logical outcome of the requirements" simply because, the requirements being in conflict, their logical outcome is an impossibility.

This is all simply to say that not even engineers can have their cake and eat it too, though they, being human, may sometimes try. To a student of the history of technology, the immobilization of New York's fleet of Grumman Flxible buses because of the occurrence of cracked steel frames was but another chapter in the Iron-Age-old story of man against manufacturing. Ever since the first ironworker—call him John Smith—tried to extract ore from the geological chaos of our terrestrial sphere and fashion it into tools, the integrity of his products has been subject to the seemingly mercurial qualities of the metal.

Swords, among the earliest iron objects, were wrought in a variety of ways to produce a keen edge and a tough blade that would not fail a warrior in battle. Yet many an ancient sword must have broken at a most unpropitious time, and the fallen warrior's comrades in arms would certainly have wanted to know why, and whether their swords would fail next. John Smith, if he wanted to stay in the sword-making business, had to apply some of his profit and experience toward research and development to come up with more reliable swords. These were no doubt easier promised than delivered, and one can easily speculate that the paucity of flawless swords raised those few apparently indestructible ones to legend. Ironically, the vastly improved reliability of modern weapons probably owes more to John Smith's understanding of why a common blade cracked or snapped in two than to his forging of an Excalibur.

The same holds true in peaceful applications of iron and steel, and the cracks in New York's buses pointed to weaknesses of design that may have inconvenienced commuters but that would be corrected in future designs to build a sounder bus frame that could take the potholes in any city's streets. Yet the nagging question arises, Does this iterative process of design by failure ever end? Will there be a day when the designers will be able to say with assurance and finality, This is a flawless design? Yes, the process can converge on a design as reliable as is reasonable; but, no, it can never be guaranteed to produce a perfectly flawless product. Design involves assumptions about the future of the object designed, and the more that future resembles the past the more accurate the assumptions are likely to be. But designed objects themselves change the future into which they will age.

It follows that departures from traditional designs are more likely than not to hold surprises. Good design minimizes the effects of surprises by anticipating troublesome details and by overdesigning for an extra measure of safety. While John Smith himself may have produced an Excalibur only by accident, the experience of failures that he could pass on to his descendants has enabled them eventually to make Excalibur a brand name and put a virtually flawless sword in every kitchen warrior's hand. This was possible because of the historical time scale over which the perfection of the sword and knife evolved. With little change in form or function from century to century, the Smiths could concentrate on the all-important metallurgy of the blade.

In contrast, it is barely two hundred years since the first iron bridge at Coalbrookdale was erected at the beginning of the Industrial Revolution. And fewer than five generations separate the introduction of the railroad train in England and the Grumman Flxible bus in New York. Unfortunately, merely knowing the history of technology does not absolve one from repeating it. While there are similarities between a nineteenth-century railway car and a late twentieth-century bus, the dissimilarities dominate.

The design of a new generation mass-transit vehicle to negotiate metropolitan traffic while satisfying guidelines for energy efficiency, accommodating the handicapped, and meeting a volume of other federal specifications is a formidable task. Though there is never any excuse for a faulty design, even under the worst constraints, there should be room for understanding. For no manufacturer wants his design to fail or endanger life. Not only is it morally wrong, it is also bad for business.

The successful transportation of men to the moon and back has demonstrated that lack of experience alone does not necessarily condemn a design to failure. It is rather the combination of inexperience, distracted by overly restrictive requirements, coupled with the pressures of deadlines, and aggravated by concerns for profit margins that initiates the cracking up of bus frames and their designers. John Smith no longer works alone. He must leave his forge regularly to go over the account books with John Doe, the supplier of his capital, and to be briefed on revised regulations by John Law, the granter of his license.

It is not only in the high technological business of building mass-transit buses that our accelerating socio-economic system breaks down. Computer models that predict the behavior of the economy have come increasingly to be relied upon to justify major economic decisions, and yet these models are not necessarily any more infallible than the ones that predict the fatigue life of a bus frame. Thus the same tools that apparently free us from the tedium of analyzing the wheel condemn us to reinvent it. We have come to be a society that is so quick to change that we have lost the benefits of one of mankind's greatest tools—experience. We are redesigning the commonest of vehicles as routinely as we are restructuring the fundamental economy. Changes are being made so radically that the relevance of lessons learned from earlier generations is not recognized. It is as if we are beating our swords into ploughshares so frenetically and carelessly as to produce blades that might fracture on striking the first pebble in their path.

Sir Alfred Pugsley, a pioneer in the study of metal fatigue in military aircraft and an articulate spokesman for the cause of structural safety, wrote: "A profession that never has accidents is unlikely to be serving its country efficiently." He was merely putting in crass terms what is a constant goal of structural engineering design in a rapidly changing society: to build safe structures more economically. The limits of structures are not always so conclusively demonstrated on their maiden flight as were the wings that fell from Icarus' arms. When an airplane flies for thousands of hours without incident, it is of course no proof that its success is a result of excessive strength that translates to more weight that in turn translates to permanent excess baggage. However, knowing that they have allowed for uncertainties in the strength of materials and uncertainties in loads on the wings and uncertainties in stress calculations with factors of safety, the engineers naturally wonder how much unnecessary weight is contained in the structure of the airplane. Thus when designing the next generation of the airplane, engineers are hard pressed to respond to questions as to why such a large factor of safety must be maintained. After all, in the intervening years they have come to understand the materials better. They have measured the loads during successful test flights of the airplane design. And they have acquired computers that enable them to make more and presumably better, more accurate stress calculations than ever before. If they do not build a lighter and more economical new plane their competitors will, and the first designers will have failed to build on their experience.

This is not unlike the way cathedrals and bridges and buses and virtually every engineering structure has evolved. With each success comes the question from society, from taxpayers, from government, from boards of directors, from engineers themselves as to how much larger, how much lighter, and how much more economically the next structure can be made. And it is not only the structural factor of safety that is skimped upon. Workmanship and style can go the way of strength. But the phenomenon is not new to the computer age. Henry Adams, forever seeing the degeneration of society, could write in *Mont-Saint-Michel and Chartres*, "The great cathedrals after 1200 show economy, and sometimes worse. The world grew cheap, as worlds must."

What happens, of course, is that success ultimately leads to failure: aesthetic failure, functional failure, and structural failure. The first can take away the zest for life, the second the quality of life, and the third life itself. Structural failure usually reverses the trend toward less and less safe structures of the kind that failed, however, either through the abandonment of that line of structures or through their being strengthened or used more conservatively. There is always pressure for relevant building codes, factors of safety, and engineering practice to be made more conservative after a failure, and in this way failures lead to new successes. The process would appear to be cyclic.

The lessons of history are clear for the structural engineer. Innovation need not be doomed to failure, for there are Iron Bridge, the Eads Bridge, and the Brooklyn Bridge—all standing as monuments to cautious exploration with new materials. There is the legacy of the Crystal Palace and there is the Empire State Building reminding us that innovative and rapid construction need not be inferior. And there is the flag on the moon to remind us that what has never been done before may be done now with the use of computers. Yet the same history tells us also that materials in new environments can lead to cracked and embrittled nuclear power plants. And it tells us that fast-track construction can leave convention-goers without convention centers. And it tells us that the computer does not necessarily make a better bus. Are the lessons of history clear for the structural engineer?

Yes, at least one lesson is clear, that innovation involves risk. Some innovations are successful and their engineers are heroes. Some innovations are failures and their engineers are goats. Yet failure in innovation should be no more opprobrious to the engineer who has prepared himself as well as he could for his attempt to build a longer bridge than to a pole vaulter who fails to make a record vault after practicing his event and using his pole to its capacity. It is the engineer who has tried to do what he is not prepared to do, or who has made the same mistake that has led to failure before, who is acting irresponsibly. He is letting down his profession as surely as the pole vaulter who refuses to practice and who uses the same kind of pole that had broken under his lighter opponent is letting down his team. The well-prepared engineer can and does build beyond experience without hubris as surely as the well-trained pole vaulter goes after a new record.

While some may wish that engineering embarrassments not be made public, it is to the credit of the profession that they are. Their being broadcast drives them home all the more so to engineers who might be working on similar computer programs, design problems, or even failure analyses. And by being not only the talk of the professional journals but also a story in the morning newspaper or on the evening news, they are less likely to be overlooked by the very person they might help the most. Engineers should not see the reports of failures as the airing of dirty laundry but rather as admissions of humanness. The journals and publications of the American Society of Civil Engineers and the British Institutions of Civil Engineers and of Structural Engineers are among the prime sources of in-depth and technical reports and analyses of structural failures because those organizations recognize the valuable lessons to be learned from failures. Their tradition is an old one, and as long ago as 1856 the celebrated Victorian civil engineer and bridge builder Robert Stephenson recommended full disclosure by authors. He wrote, in the third person, about a manuscript he had reviewed, that

> . . . he hoped that all the casualties and accidents, which had occurred during their progress, would be noticed in revising the Paper; for nothing was so instructive to the younger

Members of the Profession, as records of accidents in large
works, and of the means employed in repairing the damage.
A faithful account of those accidents, and of the means by
which the consequences were met, was really more valuable
than a description of the most successful works. The older
Engineers derived their most useful store of experience from
the observations of those casualties which had occurred to
their own and to other works, and it was most important that
they should be faithfully recorded in the archives of the
Institution.

Today Robert Stephenson would likely express the same hope,
mutatis mutandis, about the failure of computer programs and the
measures that have been taken to correct them. He would no
doubt hope that caveats would be issued whenever a potentially
dar.gerous situation in using computer-aided design is encountered, and yet he would probably be using computers himself to
design bridges beyond his own experience. But he might also have
been not a little concerned to see in the *Proceedings of the First
International Conference on Computing in Civil Engineering* only
a single session devoted to "Anatomies of Computer Disasters,"
and to read in the "abstract" of the session:

> No papers for this session will be published. The purpose
> of this is to permit the speakers to be very candid regarding
> the various computer disasters which they are describing.
> Names, organizations, etc. will not be used in order to pro-
> tect the privacy of those concerned.

While anonymity may be desirable in a litigious society, it is not
becoming of the engineering profession. And silence, rather than
being golden, here only deprives those not in attendance in New
York in 1981 of what might have been a session full of experience.
Infamous engineering failures of large structures—like the

Tacoma Narrows Bridge or the Teton Dam—are catastrophes that no one wants to see repeated. Thus it seems fitting to end a consideration of the role of failures with a paraphrase of George Santayana's familiar dictum about those not remembering, or being aware of, the past being condemned to repeat it. When I did just that in an article that appeared in *Technology Review* in 1982, I was little prepared to receive a letter from a reader who challenged Santayana's authorship of the quote so often attributed to him, and who called my attention to the fact that the familiar quotation was not in *Bartlett's*. The reader suggested that one of his ancestors and not Santayana might have first uttered the pithy saying. This surprising claim sent me on a scholarly adventure that was to demonstrate to me that humanists, like engineers, are human.

I first checked my thirteenth (1955) edition of *Bartlett's* to verify the omission of so familiar a quote, and found it to be absent. I looked in vain for what I was sure were Santayana's words in several dozen other books of quotations on my library's reference shelves before I found them in *The Reader's Digest Treasury of Modern Quotations,* along with another nice Santayana quote about welcoming the future. But the sources are given only as "George Santayana, as quoted in Reader's Digest." This was a little too insular a verification for me, so I copied down the issue dates and descended into the library's sub-basement to see if the magazine gave any further information. There, among numerous volumes of whole magazines, I began to suspect an Orwellian conspiracy when I found the pertinent pages of "Quotable Quotes" neatly excised from the bound volumes of *Reader's Digest.* Were these voids soon to be filled with tipped-in pages quoting the big brother of someone with a razor blade?

Determined to prevent the rewriting of history, I immediately returned to the reference shelves to look through every remaining book of quotations I could lay my hands on, before somebody else did. In the last one on the shelf, *The Oxford Dictionary of Quota-*

tions, I found: "Those who cannot remember the past are condemned to fulfil it." *Fulfil?* Why, this seemed to be as fantastic a discovery as finding an uncataloged folio of Shakespeare among the atlases. I would look up the reference to Santayana's *The Life of Reason* and explicate the long-misquoted text. How ironic, I thought at the time, that those who had not remembered the passage would be condemned to reread it.

The library I was using, it turns out, did not have the first (1905) edition of *The Life of Reason* but the second (1924). And it reads not "fulfil" but the pedestrian and familiar "repeat." Now I was even more excited. Perhaps Santayana had revised the word from edition to edition, and now I would be able to expound at length upon his change of heart during the Great War. And I could throw in some comments about how this change of a word in his book had escaped scholars the way a change in a detail in the Hyatt Regency walkways supports had escaped engineers.

On a steamy Saturday, on a hunch, I visited another library to look for a first edition of *The Life of Reason,* and much to my surprise I found what I imagined might be the last extant copy untouched by a razor blade. I quickly turned to page 284 in Volume One and read: "Progress, far from consisting in change, depends on retentiveness. . . . Those who cannot remember the past are condemned to repeat it." *Repeat?* I cursed *Oxford* for misleading me, and yet I thanked it for reminding me. The details, oh, the details. Why is it that we do not check them more carefully? For they can send innocent readers into the bowels of libraries, where madmen mug books, and they can send innocent citizens to their deaths in cracked airplanes and on shaky skywalks. For in an engineering design office or at a construction site, a single miscopied number or cracked part can jeopardize an entire structure and the lives that depend on its safety. That is of a lot more consequence than sending an amateur literary scholar on a wild goose chase through the stacks of libraries.

Gremlins are everywhere, in engineering design offices, in editorial offices, and in print shops. What gremlin is up to what mischief may sometimes never be possible to determine until after the fact, but by broadcasting those errors we do find we can increase our chances of catching the next anonymous glitch before it can do any harm. In that spirit I reported my experiences with the errant Santayana quote in a short essay on the op-ed page of *The Washington Post.* The piece apparently was sent out over the newspaper wire and was reprinted in at least one other paper, as I learned in a letter from a reader in Syracuse, New York. He had gone, apparently with as much incredulity as I had to my thirteenth, to his fifteenth edition of *Bartlett's* and found the troublesome quotation from Santayana printed without error. Had I had the newer edition I would never have discovered the error in *Oxford* that sent me off in all directions and taught me a lesson in details. Thus can be the luck involved in finding some errors, whether they be in books or in blueprints.

But while a missing quote can be added to new editions of *Bartlett's* or an errant word can be changed in future editions of *Oxford,* nothing can erase an engineering disaster. Yet no disaster need be repeated, for by talking and writing about the mistakes that escape us we learn from them, and by learning from them we can obviate their recurrence. As Santayana also said, "We must welcome the future, remembering that soon it will be the past; and we must respect the past, knowing that once it was all that was humanly possible."

BIBLIOGRAPHY

Adams, Henry. *Mont-Saint-Michel and Chartres.* Boston: Houghton Mifflin Company, 1933.

Addis, W. "A New Approach to the History of Structural Engineering," *History of Technology: Eighth Annual Volume, 1983,* pp. 1–13.

Alger, John R. M., and Hays, Carl V. *Creative Synthesis in Design.* Englewood Cliffs, N.J.: Prentice-Hall, 1964.

Asimow, Morris. *Introduction to Design.* Englewood Cliffs, N.J.: Prentice-Hall, 1962.

Barlay, Stephen. *The Search for Air Safety: An International Documentary Report on the Investigation of Commercial Aviation Accidents.* New York: William Morrow & Company, 1970.

Beaver, Patrick. *The Crystal Palace, 1851–1936: A Portrait of Victorian Enterprise.* London: Hugh Evelyn, Ltd., 1970.

Bettelheim, Bruno. *The Uses of Enchantment: The Meaning and Importance of Fairy Tales.* New York: Alfred A. Knopf, 1977.

Bill, Max. *Robert Maillart: Bridges and Constructions.* New York: Frederick A. Praeger, 1969.

Billington, David P. *Robert Maillart's Bridges: The Art of Engineering.* Princeton, N.J.: Princeton University Press, 1979.

———. *Structures and the Urban Environment.* Princeton, N.J.: Princeton University Department of Civil Engineering, 1983.

———. *The Tower and the Bridge: The New Art of Structural Engineering.* New York: Basic Books, 1983.

Bishop, R. E. D. *Vibration.* Cambridge: The University Press, 1965.

Blockley, D. I. *The Nature of Structural Design and Safety.* Chichester, West Sussex: Ellis Horwood Limited, 1980.

Bradley, Sculley; Beatty, Richmond Croom; and Long, E. Hudson; eds. *The American Tradition in Literature.* Revised, Shorter Edition. New York: W. W. Norton & Company, 1962.

Brown, J. Crozier. "Anatomies of Computer Disasters," *Proceedings of the First International Conference on Computers in Civil Engineering.* New York: American Society of Civil Engineers, 1981, p. 250.

Bryant, Mark. *Riddles: Ancient and Modern.* New York: Peter Bedrick Books, 1983.

Byrne, Robert. *Skyscraper.* New York: Atheneum, 1984.

Briggs, Asa. *Iron Bridge to Crystal Palace: Impact and Images of the Industrial Revolution.* London: Thames and Hudson, 1979.

Burke, John G. "Bursting Boilers and the Federal Power," *Technology and Culture,* 7 (1966), pp. 1–23.

Choate, Pat, and Walter, Susan. *America in Ruins: The Decaying Infrastructure.* Durham, N.C.: Duke University Press, 1983.

Clegg, Gordon L. *The Design of Design.* Cambridge: The University Press, 1969.

Coleman, S. N. *Bells: Their History, Legends, Making, and Uses.* Chicago: Rand McNally, 1928.

Collins, A. R., ed. *Structural Engineering—Two Centuries of British Achievement.* Chislehurst, Kent: Tarot Print Limited, 1983.

Collins, J. A. *Failure of Materials in Mechanical Design: Analysis, Prediction, Prevention.* New York: John Wiley & Sons, 1981.

Condit, Carl W. *American Building Art: The Nineteenth Century.* New York: Oxford University Press, 1960.

Cowan, Henry J. *The Master Builders: A History of Structural and Environmental Design from Ancient Egypt to the Nineteenth Century.* New York: John Wiley & Sons, 1977.

———. *Science and Building: Structural and Environmental Design in the Nineteenth and Twentieth Centuries.* New York: John Wiley & Sons, 1978.

Cowper, Charles. *The Building Erected in Hyde Park for the Great Exhibition of the Works of Industry of All Nations, 1851.* Facsimile Edition. London: Her Majesty's Stationery Office, 1971.

Cross, Hardy. *Engineers and Ivory Towers.* Edited by Robert C. Goodpasture. New York: McGraw-Hill Book Company, 1952.

Davenport, William H. and Rosenthal, Daniel, eds. *Engineering: Its Role and Function in Human Society.* New York: Pergamon Press, 1967.

Davidson, Frank P. *Macro: A Clear Vision of How Science and Technology Will Shape Our Future.* New York: William Morrow and Company, 1983.

de Havilland, Sir Geoffrey. *Sky Fever: The Autobiography.* London: Hamish Hamilton, 1961.

de Mare, Eric. *London 1851: The Year of the Great Exhibition.* London: Folio Society, 1972.

Dempster, Derek D. *The Tale of the Comet.* New York: David McKay Company, 1958.

Dixon, Roger, and Muthesius, Stefan. *Victorian Architecture.* New York: Oxford University Press, 1978.

Duga, J. J., et al. *The Economic Effects of Fracture in the United States. Part 2—A Report to NBS by Battelle Columbus Laboratories.* Washington, D. C.: U. S. Department of Commerce, 1983.

Dyson, Freeman. *Disturbing the Universe.* New York: Harper & Row, 1979.

Edwards, Llewellyn Nathaniel. *A Record of History and Evolution of Early American Bridges.* Orono, Maine: The University Press, 1959.

Emerson, Ralph Waldo. *Essays.* First and Second Series. Boston: Houghton Mifflin Company, 1925.

Evans, Ifor. *Literature and Science.* London: George Allen & Unwin, 1954.

Fairbairn, William. *On the Application of Cast and Wrought Iron to Building Purposes.* New York: John Wiley, 1854.

Feld, Jacob. *Lessons from Failures of Concrete Structures.* Detroit: American Concrete Institute, 1964.

——. *Construction Failure.* New York: John Wiley & Sons, 1968.

ffrench, Yvonne. *The Great Exhibition: 1851.* London: Harvill Press, 1950.

Flint, A. R. "Risks and Their Control in Civil Engineering," *Proceedings of the Royal Society (London) A,* 376 (1981), pp. 167–179.

Florman, Samuel C. *The Existential Pleasures of Engineering.* New York: St. Martin's Press, 1976.

——. *Blaming Technology: The Irrational Search for Scapegoats.* New York: St. Martin's Press, 1981.

Galilei, Galileo. *Dialogues Concerning Two New Sciences.* Translated by Henry Crew and Alfonso de Salvio. New York: Dover Publications. Republication of 1914 edition.

Gibbs-Smith, C. H. *The Great Exhibition of 1851.* London: Her Majesty's Stationery Office, 1950.

Giedion, Sigried. *Space, Time and Architecture: The Growth of a New Tradition.* Third Edition. Cambridge, Mass.: Harvard University Press, 1954.

Gillispie, Charles Coulston. *The Montgolfier Brothers and the Invention of Aviation, 1783–1784: With a Word on the Importance of Ballooning for the Science of Heat and the Art of Building Railroads.* Princeton, N. J.: Princeton University Press, 1983.

Godfrey, Edward. *Engineering Failures and Their Lessons.* Privately printed, 1924.

Gordon, J. E. *The New Science of Strong Materials: Or Why You Don't Fall Through the Floor.* Second Edition. Harmondsworth, Middlesex: Penguin Books, 1976.

——. *Structures: Or Why Things Don't Fall Down.* New York: Da Capo Press, 1981.

Great Britain Navy Department Advisory Committee on Structural Steels. *Brittle Fracture in Steel Structures.* London: Butterworths, 1970.

Guerber, H. A. *Myths of Greece and Rome: Narrated with Special Reference to Literature and Art.* New York: American Book Company, 1893.

Hall, Peter. *Great Planning Disasters.* Berkeley: University of California Press, 1982.

Hammond, Rolt. *Engineering Structural Failures: The Causes and Results of Failure in Modern Structures of Various Types.* New York: Philosophical Library, 1956.

Harper, Robert Francis. *The Code of Hammurabi, King of Babylon About 2250 B. C.* Second Edition. Chicago: The University of Chicago Press, 1904.

Hawthorne, Nathaniel. *The Celestial Railroad and Other Stories.* New York: New American Library, 1963.

Heins, C. P., and Firmage, D. A. *Design of Modern Steel Highway Bridges.* New York: John Wiley & Sons, 1979.

Hertzberg, R. W. *Deformation and Fracture Mechanics of Engineering Materials.* New York: John Wiley & Sons, 1976.

Hitchcock, H. R. *The Crystal Palace: The Structure, Its Antecedents and Its Immediate Progeny.* Northampton, Mass.: Smith College Museum of Art and Massachusetts Institute of Technology, 1951.

Hobhouse, Christopher. *1851 and the Crystal Palace.* London: J. Murray, 1937.

Hopkins, H. J. *A Span of Bridges: An Illustrated History.* New York: Praeger Publishers, 1970.

Janney, Jack R. *Guide to Investigation of Structural Failures.* New York: American Society of Civil Engineers, 1979.

Kemp, Emory L. "Samuel Brown: Britain's Pioneer Suspension

Bridge Builder," *History of Technology: Second Annual Volume, 1977,* pp. 1–37.

Kranzberg, Melvin, and Pursell, Carroll W., Jr. *Technology in Western Civilization.* Two Volumes. New York: Oxford University Press, 1967.

Kuhn, Thomas S. *The Structure of Scientific Revolutions.* Chicago: The University of Chicago Press, 1962.

Lathem, Edward Connery, ed. *The Poetry of Robert Frost.* New York: Holt, Rinehart and Winston, 1969.

LePatner, Barry B., and Johnson, Sidney M. *Structural and Foundation Failures: A Casebook for Architects, Engineers, and Lawyers.* New York: McGraw-Hill Book Company, 1982.

Luckhurst, Kenneth W. *The Story of Exhibitions.* London: Studio Publications, 1951.

MacGregor, J. G. "The Structural Design Process," *Introductory Report of the Twelfth Congress of the International Association for Bridge and Structural Engineering,* Vancouver, September 1984, pp. 1–12.

McKaig, Thomas K. *Building Failures: Case Studies in Construction and Design.* New York: McGraw-Hill Book Company, 1962.

Mainstone, Rowland. *Developments in Structural Form.* Cambridge, Mass.: The MIT Press, 1975.

Mark, Robert. *Experiments in Gothic Structure.* Cambridge, Mass.: The MIT Press, 1982.

Marshall, R. D., et al. *Investigation of the Kansas City Hyatt Regency Walkways Collapse.* (NBS Building Science Series 143.) Washington, D. C.: U. S. Department of Commerce, National Bureau of Standards, 1982.

McCullough, David. *The Great Bridge.* New York: Simon and Schuster, 1972.

Mock, Elizabeth B. *The Architecture of Bridges.* New York: The Museum of Modern Art, 1949.

Morse, Samuel French, ed. *Poems of Wallace Stevens.* New York: Vintage Books, 1959.

Nervi, Pier Luigi. *Structures.* Translated by Giuseppina and Mario Salvadori. New York: F. W. Dodge Corporation, 1956.

Orenstein, Glenn S. "Instant Expertise: A Danger of Small Computers," *Civil Engineering,* 54 (June 1984), pp. 50–51.

Osgood, Carl C. *Fatigue Design.* Second Edition. Oxford: Pergamon Press, 1982.

Owen, N. B., Turner, C. E., and Irving, P. E. "The Failure of the Quarter Bells Chiming Mechanism of the Great Clock, Palace of Westminster," *Big Ben–Its Engineering Past and Future.* London: The Institution of Mechanical Engineers, 1981, pp. B1–B28.

Perrow, Charles. *Normal Accidents: Living with High-Risk Technologies.* New York: Basic Books, 1984.

Petroski, Henry. "Reflections on a Slide Rule," *Technology Review,* 84 (February/March 1981), pp. 34–35.

——. "When Cracks Become Breakthroughs," *Technology Review,* 85 (August/September 1982), pp. 18–30.

——. "On 19th Century Perceptions of Iron Bridge Failures," *Technology and Culture,* 24 (1983), pp. 655–659.

——. "The Details, Oh, the Details," *The Washington Post,* November 10, 1982, Sect. A, p. 31.

———. "The Amazing Crystal Palace," *Technology Review,* 86 (July 1983), pp. 18–28.

———. "Offshore Engineering: Oil from Troubled Waters," *Technology Review,* 87 (July 1984), pp. 52–61, 76.

Plowden, David. *Bridges: The Spans of North America.* New York: The Viking Press, 1974.

Provenzo, Eugene F., Jr., and Brett, Arlene. *The Complete Block Book.* Syracuse, N. Y.: Syracuse University Press, 1983.

Pugsley, Sir Alfred. *The Safety of Structures.* London: Edward Arnold, 1966.

Pye, David. *The Nature and Aesthetics of Design.* New York: Van Nostrand Reinhold Company, 1978.

Rosenfield, A. R. "The Crack in the Liberty Bell," *International Journal of Fracture,* 12 (1976), pp. 791–797.

Ross, Steven S. *Construction Disasters: Design Failures, Causes, and Prevention.* New York: McGraw-Hill Book Company, 1984.

Salvadori, Mario. *Why Buildings Stand Up: The Strength of Architecture.* New York: McGraw-Hill Book Company, 1982.

Salvadori, Mario, and Tempel, Michael. *Architecture and Engineering: An Illustrated Teacher's Manual on Why Buildings Stand Up.* New York: The New York Academy of Sciences, 1983.

Sandstrom, Gosta E. *Man the Builder.* New York: McGraw-Hill Book Company, 1970.

Schmid, Herman, and Busch, David S. "An Electronic Digital Slide Rule," *The Electronic Engineer,* 27 (July 1968), pp. 54–64.

Scott, Quinta, and Miller, Howard S. *The Eads Bridge.* Columbia: University of Missouri Press, 1979.

Selected Papers from the First National Conference on Civil Engineering: History, Heritage, and the Humanities. Two volumes. Princeton, N.J.: Princeton University, 1970.

Shute, Nevil. *No Highway.* New York: William Morrow & Co., 1948.

Silby, P. G., and Walker, A. C. "Structural Accidents and Their Causes," *Proceedings of the Institution of Civil Engineers,* Part 1, 62 (May 1977), pp. 191–208.

Smith, Cyril Stanley. *A History of Metallography: The Development of Ideas on the Structure of Metals before 1890.* Chicago: The University of Chicago Press, 1960.

Smith, Denis. "The Use of Models in Nineteenth Century British Suspension Bridge Design," *History of Technology: Second Annual Volume, 1977,* pp. 169–214.

Smith, H. Shirley. *The World's Great Bridges.* New York: Harper & Brothers, 1953.

Smith, Julian. *Nevil Shute (Nevil Shute Norway).* Boston: Twayne Publishers, 1976.

Sporn, Philip. *Foundations of Engineering.* New York: The Macmillan Company, 1964.

Steinman, D. B. *The Builders of the Bridge: The Story of John Roebling and His Son.* Second Edition. New York: Harcourt, Brace and Company, 1950.

———. *A Practical Treatise on Suspension Bridges: Their Design, Construction and Erection.* Second Edition. New York: John Wiley & Sons, 1929.

Straub, Hans. *A History of Civil Engineering: An Outline from Ancient to Modern Times.* Translated by E. Rockwell. London: Leonard Hill Limited, 1952.

Talese, Gay. *The Bridge.* New York: Harper & Row, 1964.

Tedesko, Anton. "How Have Concrete Shell Structures Performed? An Engineer Looks Back at Years of Experience with Shells," *Bulletin of the International Association for Shell and Spatial Structures,* 21 (August 1980), pp. 3–13.

Tedesko, A., and Billington, D. P. "The Engineer's Personality and the Influence It Has on His Work—A Historical Perspective," *Concrete International: Design and Construction,* 4 (December 1982), pp. 20–26.

Thompson, D'Arcy Wentworth. *On Growth and Form.* New Edition. Cambridge: The University Press, 1944.

Timoshenko, Stephen P. *History of Strength of Materials: With a Brief Account of the History of Theory of Elasticity and Theory of Structures.* New York: McGraw-Hill Book Company, 1953.

Torroja, Eduardo. *Philosophy of Structures.* Translated by J. J. Polivka and Milos Polivka. Berkeley: University of California Press, 1962.

——. *The Structures of Eduardo Torroja: An Autobiography of Engineering Accomplishment.* New York: F. W. Dodge Corporation, 1958.

Trachtenberg, Alan. *Brooklyn Bridge: Fact and Symbol.* New York: Oxford University Press, 1965.

U. S. House of Representatives Committee on Science and Technology. *Structural Failures: Hearings Before the Subcommittee on Investigations and Oversight.* Washington, D. C.: Government Printing Office, 1983.

———. *Structural Failures in Public Facilities.* Washington, D. C.: Government Printing Office, 1984.

Vannoy, Donald W. "20/20 Hindsight with the Aid of a Computer," *Professional Engineer,* 53 (Winter 1983), pp. 21–25.

Vose, George L. *Bridge Disasters in America: The Cause and the Remedy.* Boston: Lee and Shepard, 1887.

Watson, Wilbur J., and Watson, Sara Ruth. *Bridges in History and Legend.* Cleveland, Ohio: J. H. Jansen, 1937.

Werner, Ernst. *Der Kristallpalast zu London 1851.* Dusseldorf: Werner-Verlag, 1970.

Whipple, S. *An Elementary and Practical Treatise on Bridge Building.* Fourth Edition. New York: D. Van Nostrand, 1883.

Whyte, R. R., ed. *Engineering Progress through Trouble.* London: The Institution of Mechanical Engineers, 1975.

Zetlin, Lev, Associates. *Report of the Engineering Investigation Concerning the Causes of the Collapse of the Hartford Coliseum Space Truss Roof on January 18, 1978.* New York: Lev Zetlin Associates, 1978.

Zuk, William. *Concepts of Structure.* New York: Reinhold Publishing Corporation, 1963.

In addition, *Punch,* the *Illustrated London News,* and other nineteenth-century periodicals have provided historical points of reference, and the following are a continuing source of relevant items and information: *Civil Engineering, Engineering News-Record, The New York Times, The Structural Engineer,* and *Technology and Culture.*

INDEX

Adams, Henry, 55, 222
Addison, Joseph, 70–72
Advisory Committee on Reactor
    Safeguards (ACRS), 114–115
Air Registration Board, 179
Aircrash Detective (Barlay), 184–186
airline crashes, 5, 176–184
    see also DC-10 crash
Airspeed, Ltd., 181
Albert, Prince Consort of England,
    139, 143
Alexander L. Kielland, 172–176, 187
    design and use of, 172–173
    reasons for collapse of, 173–175
alternate load paths, 92–93
American Iron and Steel Institute,
    130
American Society of Civil Engineers,
    70, 202–203, 211, 223
Ammann, Othmar, 76, 163–164
aqueducts, Roman, 67
Architecture and Engineering
    Performance Information Center
    (AEPIC), 210–212
art, engineering as, 40, 80
Ashtabula, Ohio, railroad bridge
    collapse at, 70

Atomic Energy Commission, 114
Auth, Tony, 3
automobile accidents, 5

Barlay, Stephen, 184–186
Barlow, Peter, 140
Bartlett's Familiar Quotations, 225,
    227
Battelle Columbus Laboratories, 7
beams:
    bending of, 46, 50
    buckling of, 47
    cantilever, 49–51
    definition of, 45
    strength of, 50
Beauvais, cathedral at, 55–56
Bent pyramid, 55
Bettelheim, Bruno, 18
Big Ben, 108–109
Billington, David, 83, 101
Blockley, D. I., 207–208
Board of Trade (England), 69
Boeing Company, 180
bridges:
    early failures of, 56, 58, 59
    history of, 67–72
    modern failures of, 93–95

bridges *(continued)*
  symbolic nature of, 57
  see also railroad bridges;
    suspension bridges; *specific
    bridges*
Britannia Bridge, 101, 140, 162
British Institution of Civil Engineers,
  150, 223
British Institution of Structural
  Engineers, 40, 223
brittle fractures, 109, 115–116
Bronx-Whitestone Bridge, 163, 165
Brooklyn Bridge, 76
  design and construction of,
    158–163
  structural soundness of, 45, 162,
    170
  symbolic value of, 57, 158, 160,
    162, 171, 222
Brunnel, Isambard Kingdom, 101
building codes, 86–87
Building Failures (McKaig),
  204–207, 211
Bunyan, John, 59
Byrne, Robert, 212

calculators, 191–193
California Institute of Technology,
  166
Canadian Concrete Code Committee,
  201
cantilever beams, 49–51
cathedrals, medieval, 55–56, 57, 82
"Celestial Railroad, The"
  (Hawthorne), 59–61
Centre d'Art et de Culture
  Georges-Pompidou, 153
Chicago blizzard (1979), 49
China Syndrome, The, 117
Cincinnati Bridge, 162, 163

Coalbrookdale, iron bridge at,
  67–68, 219
Cole, Henry, 138
Commodore, 191
computer-aided design (CAD), 195
computers:
  engineering impact of, 194
  misuses of, 193, 195, 198–203
  shortcomings of, 196–197
  testing of, 199–200
"Connoisseur of Chaos" (Stevens),
  81
Construction Disasters (Ross), 212
"Contagiousness of Puerperal Fever,
  The" (Holmes), 28
Crystal Palace, 101, 136–157, 160,
  222
  construction of, 141–142, 143
  decoration of, 146–147
  design of, 140, 142–143
  destruction of, 150
  inspiration for, 137
  modern architecture influenced by,
    149, 151–157
  safety concerns about, 143–146
  Sydenham reconstruction of,
    149–150
Crystal Palace Engineering School,
  150

Daedalus, 185–186, 216–217
Dahshur, pyramid at, 55
Dallas Market Center, 153
Daniel Guggenheim Aeronautical
  Laboratory, 166
DC-10 crash (Chicago), 1, 5, 26, 105
  causes of, 95–96, 112, 187, 208
  effects of, 72, 96
"Deacon's Masterpiece, The"
  (Holmes), 29, 35–39, 212

de Havilland, Geoffrey, 150, 180
de Havilland Aircraft Company, 181
de Havilland Comet, 176–184, 187,
    204, 208
    crashes of, 176–177
    final success of, 180
    reasons for failure of, 178–180
Denver, paraboloidal shell at, 83–84
design:
    choices in, 66, 73–74, 80
    compromise as part of, 216–219
    as engineering, xi
    experience in, 66–67, 104, 105,
        220
    factor of safety in, 98
    fatigue and, 24–26
    and lifetime of structure, 30–31
    resisting forces in, 41
    revision in, 75, 78, 83–84
    risk in, 107, 222
    role of failure in, xii, 62, 79,
        82–84, 97, 105, 121, 180, 206,
        219, 227
    scientific hypotheses and, 43–45,
        62
    trial and error in, 53–59, 62–63
design errors, in building failures,
    208–209
*Dialogues Concerning Two New
    Sciences* (Galileo), 49
Dixon, Ill., railroad bridge collapse
    at, 70
Duke University, xii–xiii, 84
Dum-Dum Airport, 176

Eads, James, 159
Eads Bridge, 158–160, 222
Edison, Thomas, 27
Eiffel Tower, 158, 171
Einstein, Albert, 43

Electric Power Research Institute,
    199
"Electronic Digital Slide Rule, An,"
    191
Emerson, Ralph Waldo, 58
Empire State Building, 156, 222
engineering:
    as art, 40, 80
    as design, xi
    forensic, 176–188
    as human nature, xi, 11–20
    purpose of, 9
    as science, 40, 49, 62, 80
Engineering News-Record, 89–90,
    166, 212
engineers:
    creativity of, 79–80
    definitions of, 61
    fallibility of, 52
    as poets, 81–83
    responsibilities of, 52
Escher, M. C., 89

factor of safety, 114
    calculation of, 99–102
    in Hyatt Regency Hotel, 102, 114
    lowering of, 200
    purpose of, 98–99
failures, engineering:
    alternate load paths in, 92–93
    analysis of, 174–188
    anticipation of, 27, 187, 206, 208,
        216
    causes of, 204–205, 207–209
    costs of, 7–8
    cyclical nature of, 101, 222
    design errors in, 208–209
    disproved hypotheses and, 44–45
    dissemination of information and,
        209–212, 214, 223, 227

failures *(continued)*
   factor of safety and, 100, 103–104
   by fatigue, 21–29
   human error in, 204–205,
      207–209, 213–214
   inevitability of, 9–10, 28–30,
      187–188
   learning from, xii, 53–59, 62–63,
      76, 79, 82–84, 97, 105–106, 121,
      180, 223–224, 227
   in mass-produced vs. unique
      objects, 26
   prevention of, 206, 209
   probability of, 4–5, 29, 44, 45, 73
Fairbairn, William, 162
Farnborough Royal Aircraft
      Establishment, 177, 183, 185
fatigue:
   design and, 24–26
   in iron railroad bridges, 68–69
   probability of, 21–25, 28, 29
fatigue cracks, 107–135
   analysis of, 126–135
   examples of, 109, 124, 126, 174–
      175, 179–180
   mechanism of, 109–110
   in nuclear power plants, 114–
      118
   prevention of, 110–114
   testing for, 111–113
Fatigue Design (Osgood), 109
Federal Aviation Administration,
      5
Federal Urban Mass Transit
      Administration, 123
Finnegans Wake (Joyce), 78
Fox, Charles, 101
Fox, Henderson, and Co., Messrs.,
      141, 145

Frost, Robert, 74
Futures Group, 192

Galileo, 49–52
Galloping Gertie, *see* Tacoma
      Narrows Bridge
George Washington Bridge, 163
Gleason, Jackie, 124
Gore, Albert, 210
Great Conservatory (Chatsworth),
      137
Great Exhibition, *see* Crystal Palace
Great Palm House (Kew Gardens),
      137
Great Western Railway, 101
Greeley, Horace, 147
Growald, Martin, 153–154
Grumman Flxible buses, 123–126,
      134, 218, 219

Hammurabi, Code of, 3–4
Harper's Weekly, 60, 70
Hartford Civic Center, 198–199,
      200
Harvard Medical School, 28
Hawthorne, Nathaniel, 59–61, 72
Hennebique, François, 82
Hershey, Pa., arena at, 83
Holmes, Oliver Wendell, 28–29,
      35–39, 212
Hoover, Herbert, 214–215
House Committee on Science and
      Technology, 209
Household Words, 151
human body, as engineering feat,
      11–20
human error, in building failures,
      204–205, 207–209, 213–214
Huxley, T. H., 84

Hyatt Regency Hotel (Chicago), 92
Hyatt Regency Hotel (Kansas City),
  1, 26, 44, 72, 85–93, 125, 161,
  187, 204, 208
  building code violations of, 4, 86–
  87
  casualties of, 85
  factor of safety in, 102, 114
  legal actions against, 4, 91
  skywalk design of, 86–92, 95,
  102–103

IBM Building (New York), 153
Icarus, 184–186, 216–217
Illustrated London News, 60, 139,
  140, 145
Imhotep, 54
Infomart, 153–154
infrastructure, neglect of, 7–8
International Association for Bridge
  and Structural Engineering, 201
iron, in bridges, 56, 68–72
Ironbridge (Coalbrookdale), 67–68,
  222

John Hancock Building, 210
Jones, Owen, 146–147, 152
Joyce, James, 77–78

Kansas City Star, 86
Kármán, Theodore von, 166
Keuffel & Esser (K & E), 189–193

LePatner, Barry, 210
Lever House, 152
Liberty Bell, 107–108, 111
Liberty ships, 115–116
Life of Reason, The (Santayana),
  226

McDonnell-Douglas, 176
MacGregor, James G., 201–202
McKaig, Thomas, 204–207, 208, 211
Mackinac Bridge, 169
magnetic fields, for parts testing, 113
Maillart, Robert, 74, 81–83
Maine River Bridge, 160
*Man Who Could See Through Time,
  The* (Wagener), 76
Mark, Robert, 56
Marshall, John, 108
Maryland, University of, 210
mastabas, 54
Masterpiece Theatre, 181
Mead Prize, 202–203
Meidum pyramid, 55
Menai Straits Suspension Bridge, 168
Mianus Bridge, 93–94, 95
Micro-Measurements, 29
Mies van der Rohe, Ludwig, 152–153
Mont-Saint-Michel and Chartres
  (Adams), 222
Murphy's Law, 28, 159

National Bureau of Standards, 7, 86,
  88, 102, 209–210
National Science Foundation, 211
National Transportation Safety
  Board, 94, 210
*Nature of Structural Design and
  Safety, The* (Blockley), 207
Newton, Sir Isaac, 43
New York City, water main break in
  (1983), 7–8
New York City Marathon, 122–123
New York City Transit Authority
  (NYCTA), 124–125, 134
New York Convention Center,
  154–156

New York Times, The 76, 125
Niagara Bridge, 162–163
No Highway (Shute), 181–184, 212
nuclear power plants:
  fatigue cracks in, 115–117
  safety factors in, 119–120
  stress-corrosion cracking in,
    118–119
Nuclear Regulatory Commission,
  118, 197

offshore oil rigs, 172–175
On Medical Education (Huxley), 84
On the Beach (Shute), 181
"On the Projected Kendal and
  Windermere Railway"
  (Wordsworth), 58–59
Osgood, Carl, 109
Otis, Elisha, 149
overdesign, 73, 101, 111, 194
Ovid, 185–186
*Oxford Dictionary of Quotations,
  The,* 225–227
Oxford English Dictionary, 61

Palma, Majorca, cathedral at, 56
Paxton, Joseph, 136–137, 140–143,
  147, 149–151
Pei, I. M., 83
Pilgrim's Progress, The (Bunyan),
  59
Point Pleasant Bridge, 94–95, 105
Prospect Tower, 148
Pugsley, Sir Alfred, 183, 221
Pulitzer Prize, 86
Punch, 60, 70–72, 144, 146
Pye, David, 217
pyramids, 53–55, 57

Quebec Bridge, 69

railroad bridges:
  collapses of, 56–59, 68–73, 138, 144
  iron in, 56, 68–72
  public responses to, 58–61, 70–72
Reader's Digest, 225
*Reader's Digest Treasury of Modern
  Quotations,* 225
reference temperatures, 115, 116–117
relativity theory, 43
resonance, structural, 161
Rich, Frank, 76
Roberts-Austen, Sir William
  Chandler, 130
Rockwell, Norman, 31
Roebling, Emily, 169
Roebling, John, 159, 162–163, 169
Roebling, Washington, 76, 159, 161,
  169
Ross, Steven S., 212
Royal Aeronautical Society, 181
Royal Botanic Gardens, 137
Royal Gold Medal in Architecture,
  150–151
Ruined City (Shute), 181
Ruskin, John, 138

safe-life criterion, 114
safety, 6, 41
  see also factor of safety
St. Louis, airport at, 83–84
St. Paul's Cathedral, 139
Salazar, Alberto, 122–123
Santayana, George, 225–227
Savalas, Telly, 124
science, engineering as, 40, 49, 62, 80
scientific hypotheses:
  creation of, 42
  design and, 43–45, 62
  testing of, 42–45, 47–48, 51, 80,
    104–105

scientific method, 42, 62
Seagram Building, 152–153
Sears Catalog, 31
Severn Bridge, 169–170
Shute, Nevil, 181–184, 212
Sibthorp, Charles, 139–140, 148
significant digits, 190
Silver Bridge, 94–95
Skylab, 3
Skyscraper (Byrne), 212–214
slide rules, 189–193
Smithsonian Institution, 192
Society for the History of
Technology, 69
Society of Arts (England), 138
Speak and Spell, 22–27
Spectator, 70
stainless steel, 130–131, 133
"Steamboats, Viaducts and
Railways" (Wordsworth), 59
steel, 56, 83
Steinman, David, 169
Stephenson, Robert, 101, 140,
162–163, 223
Stevens, Wallace, 75, 81
stress-corrosion cracking, 133
in nuclear power plants, 118–119
Structural Engineer, 40
suspension bridges:
aerodynamic instability of, 164–166
collapses of, 164–168
history of, 158–163
Sydenham Crystal Palace, 149–150

Tacoma Narrows Bridge, 1, 3, 44–
45, 69, 80, 125, 162, 225
design of, 164
reasons for collapse of, 164–168
Tay Bridge, 69

Technology and Culture, 69
Technology Review, 23, 90, 225
Tedesko, Anton, 83–84
Telford, Thomas, 168
Tenniel, John, 70, 72, 146
Teton Dam, 45, 225
Texas, University of, at Austin, 192
Texas Instruments, 22, 191–192
Three Mile Island accident, 1, 3,
197
Times (London), 139
Town Like Alice, A (Shute), 181
Trammel Crow Company, 153
Transmission Line Mechanical
Research Facility, 199

ultrasonic waves, for parts testing,
112–113
Ulysses (Joyce), 78

Verrazano Narrows Bridge, 76, 122
Victoria, Queen of England, 68, 73,
144–145, 147–148
Vienne River, bridge on, 82–83
"Vision of Mirzah, The" (Addison),
70–72

Wagener, Terri, 76
Washington, George, 108
Washington Monument, 101–102
Washington Post, 227
Wheeling *Intelligencer,* 167
Wheeling Suspension Bridge, 166
Wordsworth, William, 58–59, 60

X-rays, for parts testing, 113

Zoser, 54
Zuoz, bridge at, 83